Blood and Guts

ROY PORTER

Blood and Guts

A SHORT HISTORY OF MEDICINE

W. W. Norton & Company
New York • London

First published in Great Britain 2002 by Allen Lane,
an imprint of Penguin Books

First published as a Norton paperback 2004

Printed in the United States of America

Manufacturing by The Haddon Craftsmen, Inc.
Production manager: Anna Oler

Library of Congress Cataloging-in-Publication Data

Porter, Roy, 1946–
Blood and guts : a short history of medicine / Roy Porter.—1st American ed.
p. cm.
Includes bibliographical references and index.
ISBN 0-393-03762-2
1. Medicine—History—Popular works. I. Title.

R131.P587 2003
610'.9—dc21 2003042078
ISBN 0-393-32569-5 pbk.

W. W. Norton & Company, Inc., 500 Fifth Avenue, New York, N.Y. 10110
www.wwnorton.com

W. W. Norton & Company Ltd., Castle House,
75/76 Wells Street, London W1T 3QT

567890

To Natsu the panacea

Contents

List of Illustrations

The author and publishers would like to express their thanks to the following sources of illustrative material and for permission to reproduce it: The Punch Library, title page, 13 and 38; The National Library of Medicine USA, 4, 19, 22, 23 and 35; The Bodleian Library, University of Oxford (MS Ashmole 1462, fol. 9v), 26; © The British Library London, 27; © Tate London, 2002, 12. All other images are courtesy of the Wellcome Library, London.

Acknowledgements

This book originated in lecture courses given for many years at the Wellcome Institute (reborn since October 2000 as The Wellcome Trust Centre for the History of Medicine at University College London), and some of it was written there before my retirement in September 2001. For making this book possible, I wish to acknowledge the enormous support given to me over the years by many members of the Institute's staff, notably my secretaries Frieda Houser and later Rebecca Baker and Emma Ford, and the administrator Alan Shiel. My gratitude to Alan and to the Director, Hal Cook, for making such a generous contribution to the cost of the illustrations. Not least, thanks to all the students who made my twenty years at the Institute such a pleasure and whose critical responses helped me develop my ideas. I hope future students will find this book as stimulating as I found those I taught.

Drafts have been read by Hal Cook and Natsu Hattori, to whom I am deeply grateful for their customarily candid criticism and shrewd suggestions. Retyping of numerous versions has been done by Laurent Busy, Caroline Coulter, Debra Scallan and the indefatigable Sheila Lawler. Thanks also to Jed Lawler for coming yet again to the rescue of a computer buffoon.

The enthusiasm of Simon Winder at Penguin has, as always, been a delight. Bela Cunha proved an excellent copy-editor and Jane Henderson compiled the index with her usual acumen.

The art has three factors, the disease, the patient, the physician. The physician is the servant of the art. The patient must cooperate with the physician in combating the disease.

Hippocrates, *Epidemics*, *I*, 11

Physician, heal thyself.

St Luke, 4: 23

Preface

This survey explores the historical interaction of people, disease and health care, set in context of societies and their beliefs. Brevity compels me to narrow my focus to Western medicine – unique in being the only tradition which has succeeded in globalizing itself. Emphasizing change over continuity, I tell my story with as much narrative detail as space allows: disease (Chapter 1); healers in their various incarnations (Chapter 2); the investigation of the body (Chapter 3); the modern biomedical sciences pioneered in the laboratory and the consequent biomedical model of disease (Chapter 4); therapeutics, especially in the scientific age (Chapter 5); surgery (Chapter 6); and that key medical institution, the hospital (Chapter 7). The concluding discussion (Chapter 8) assesses the wider socio-political aspects and implications of modern medicine.

All too little will be said of the personal side: how people have experienced sickness and how it has affected their lives. But sufferers' responses to being ill or incapacitated, and to the threat of dying loom over this book as an ever-present shroud. The dread of disease, potential and actual, the pains of acute complaints and long-term ailments, and the terror of mortality number among our most universal and formidable experiences. Religion and philosophy are arguably the products of mankind's efforts to cope, in mind and heart, individually and collectively, with afflictions and death.

In countless ways, through multitudes of folk precepts and practices, societies have sought to hold disease at bay, or to fight, manage and rationalize it when it occurs. In answer to that nagging question 'why me?', maladies have often been personified, held as judgements, or given moral meanings. Thus there are 'bad' diseases (leprosy, that biblical plague, or syphilis – both of which carried stigmas, as in 'moral leper') but also 'good' ones (tuberculosis, for instance, so often linked with Romantic genius, or gout, the insignia of a gentleman). Disease may also be read as the wrath of God – an archaic idea which surfaced again with AIDS. Medical anthropologists have shown how beliefs about the body, in sickness and in health, are central to social value systems, indeed to what has been called the 'body politic'.

Focusing as it does on the history of medicine, this book cannot delve into these personal inflections of sickness and the experience of 'embodiment' – to follow up such issues, please consult the Further Reading section. Anxieties about disease and doctors have, however, been omnipresent. And if, as we must, we regard the self as a mind/body continuum and sickness as in part psychosomatic, such fears must not be viewed as peripheral but as integral to the story of suffering and its relief. The agonies of the sick and the dying haunt the story of disease and medicine told below.

CHAPTER ONE

Disease

> And I looked, and behold a pale horse and his name that sat on him was Death, and Hell followed with him. And power was given unto them over the fourth part of the earth, to kill with sword, and with hunger, and with death, and with the beasts of the earth.
>
> The Book of Revelation, 6:8

The war between disease and doctors fought out on the battleground of the flesh has a beginning and a middle but no end. The history of medicine, in other words, is far from a simple tale of triumphant progress. As is hinted by the story of Pandora's box or the Christian Fall, plagues and pestilences are more than inevitable natural hazards which can, we hope, be overcome: they are largely of mankind's own making. Epidemics arose with society, and sickness has been, and will remain, a social product no less than the medicine which opposes it. Civilization brings not just discontents but diseases.

Some five million years ago, anthropologists tell us, Africa witnessed the first ape men, the low-browed, big-jawed Australopithecines. Within three million years, there had evolved our upright, large-brained ancestor *Homo erectus*, who learned how to make fire, to use stone tools, and (eventually) to speak. This omnivore fanned out about a million years ago into Asia and Europe, and a direct line leads, around 150,000 BC, to *Homo sapiens*.

Hunter-gatherers beset by harsh and dangerous environments, our palaeolithic precursors led brief lives. Nevertheless,

1. Death Sitting on a Globe. *Frontispiece from English Dance of Death, Thomas Rowlandson, 1816.*

they escaped the plagues that were to besiege later societies. Somewhat like the bush-people of the Kalahari, they were nomads who lived in small and scattered groups. Infectious diseases (smallpox, measles, flu and the like) must have been virtually unknown, since the micro-organisms responsible for them require high population densities to provide reservoirs of susceptibles. Neither did these isolated hunter-foragers stay put long enough to pollute water sources or deposit the filth which attracts disease-spreading insects. Above all, they lacked the tamed animals which have played so equivocal a role in human history. While domesticated creatures have made civilization possible, they have also proved continual and often devastating sources of sickness.

As humans colonized the globe, they were themselves colonized by pathogens. These included parasitic worms and insects – helminths, fleas, ticks and arthropods; and also micro-organisms such as bacteria, viruses and protozoans, whose ultra-rapid reproduction rates produce severe illnesses within a host but generally – the silver lining – provoke in survivors some immunity against reinfection. Such microscopic enemies became locked with humans in evolutionary struggles for survival characterized not by ultimate winners and losers but by uneasy coexistence.

As it multiplied, the human race moved out of Africa, first into the warm regions of Asia and southern Europe, and then further north. Nomadic ways continued until the end of the last Ice Age (the Pleistocene) some 12,000–10,000 years ago. With the depletion of game supplies and no further huge virgin tracts of land available which were rich in game, population pressure drove mankind to tilling the soil – there was no alternative, it was a matter of produce or perish.

Faced by famine, humans learned by trial and error to harness natural resources and grow their own food. They began to breed wild grasses into domesticated grains – wheat, barley, rice, maize, etc. – and to bring dogs, cattle, sheep, goats, pigs, horses and poultry under control. Within just a few thousand years Stone Age stalkers thereby turned into pastoralists and tillers of the soil, able to dominate their less advanced neighbours. Mankind passed its first survival test.

With its animal husbandry and systematic agriculture, settlement allowed populations to spiral. Clearing woodland, harvesting crops and preparing food were all labour-intensive activities, and so required more hands – who could now in turn be fed. Such developments in time brought the formation of more organized and permanent communities (villages, towns, cities) with their chiefs, laws and social hierarchies, and later their courts and officials. Among other trades and statuses, there arose specialist healers.

But if the coming of agriculture delivered mankind from the Malthusian threat of starvation, it also unleashed a new danger: infectious disease. For pathogens which had once been exclusive to animals became transferred, through long and complex evolutionary processes, to humans: animal diseases leapt the species gap and mutated into human ones. In the course of history such Darwinian adaptations have led to a situation in which humans nowadays share more than sixty micro-organic diseases with dogs, and only slightly fewer with cattle, sheep, goats, pigs, horses and poultry.

Back in Neolithic times cattle contributed tuberculosis, smallpox and other viruses to the human pathogen pool. Pigs and ducks passed on their influenzas, while horses brought rhinoviruses, not least the common cold. Measles is the result of rinderpest (canine distemper) jumping to humans from dogs and cattle. A recent instance of such developments is today's BSE/CJD crisis – bovine spongiform encephalopathy in livestock is the source of human Creutzfeldt-Jakob disease. Greedy and slovenly farming practices will facilitate the leap of further new diseases from animals to humans.

And humans proved vulnerable in other ways. Farm and domestic animals and vermin carry salmonella and other bacteria; faecally polluted water spreads polio, cholera, typhoid, hepatitis, whooping cough and diphtheria; and granaries become infested with bacteria, toxic fungi, rodent excrement and insects. Settlements, in short, bade disease to settle too.

Meanwhile worms took up permanent residence within the human body. The parasitic roundworm *Ascaris* probably evolved in humans from pig ascarids, leading to diarrhoea and malnutrition. Other wormlike helminths colonized the gut, including the yards-long hookworm and the filarial worms responsible for tropical elephantiasis and river blindness. Severe diseases became endemic wherever agriculture depended upon irrigation – in Mesopotamia, Egypt, India and around the great rivers of southern China. Paddy-fields harbour parasites which enter the bloodstream of barefoot workers, including the blood fluke *Schistosoma* that causes bilharzia or schistosomiasis ('big belly').

Permanent settlement thus afforded golden opportunities for insects, vermin and parasites. Moreover, agriculture led to excessive reliance on starchy monocultures such as maize, low in proteins, vitamins and minerals. Stunted people are more prone to sickness, and poor nutritional levels in turn led to pellagra, marasmus, kwashiorkor, scurvy and other deficiency diseases. In the transition from nomadic to Neolithic society, the scales of health tipped unfavourably, with infections worsening and vitality declining. People grew shorter.

Settlement also brought malaria, to this day a scourge in warm climates and perhaps, with global warming, about to spread still further. First in sub-Saharan Africa, conversion of forests into farmland created the warm water-holes and furrows which make perfect breeding environments for mosquitoes. The symptoms of malarial fevers were familiar to the Greeks but were not scientifically explained until around 1900, when the new tropical medicine showed how they are produced by the microscopic protozoan parasite *Plasmodium*, which lives in the *Anopheles* mosquito. Transmitted to humans through its bites, the parasites then move through the bloodstream to the liver, where they breed during an incubation stage of a couple of weeks. Returning to the blood, they then attack the red blood cells, whose breakdown causes recurrent violent chills and high fever.

Haunting agricultural settlements, malaria moved out from Africa (to this day heavily malarial) to the Near and Middle East and Mediterranean. India also proved ripe for infection, as did China's southern coastal strip. And from the sixteenth century Europeans shipped it to the New World.

Despite such rampant infections bred by congested settlement rife with waste and dirt, mankind's ambitions and restless energies ensured that, however unhealthy, communities expanded. More humans spawned more diseases in explosive surges which were temporarily checked but never terminated. Before the invention of agriculture, global population may have been around the five million mark; by 500 BC, the golden age of Athens, it had leapt to perhaps 100 million; by the second century AD it may have doubled, while the 2000 figure was some 6,080 million, with projections of a further doubling within the next century.

Renewed population pressure brought widespread privation and more meagre diets. But however malnourished, parasite-riddled and pestilence-battered, the human race did not prove totally defenceless against the onslaught of disease. Survivors of epidemics acquire some antibody protection; and in the long run the survival of the fittest means that immune systems grow more sophisticated, enabling humans to co-exist with their micro-organic foes. Immunities passed via the placenta or mother's milk provide infants with some defence, and genetic shields have been developed, as with the sickle-cell trait among black Africans which protects against vivax malaria (this, ironically, made them ideal workers on New World slave plantations). Darwinian adaptations might thus take the edge off lethal afflictions.

But the threats remained dire, especially to virgin populations. By 3000 BC great city-empires were rising in Mesopotamia and Egypt, in the Indus Valley and on the Yellow River; Mesoamerica followed. In the Old World, such settlements maintained huge cattle herds, from which lethal pathogens, notably smallpox, spread to humans. Other originally zoognostic (animal-based) conditions – diphtheria, influenza, chickenpox, mumps and so forth – also began to have a devastating impact on packed populations which as yet had no immunity. Unlike malaria, these zoonoses need no carriers and, being directly contagious, they spread readily and rapidly.

Thus began the era of catastrophic epidemics. With the incessant outreach of civilization and exchange of goods, merchants, mariners and marauders fatally bestowed on untouched susceptibles the Greek gift of disease. One region's familiar 'tamed' disease would become another's deadly plague as trade, travel and war detonated pathological explosions. In the transmission of diseases, the city's role was decisive. Until recent times, towns were so insanitary and pest-ridden that their populations never replaced themselves naturally. Their multiplication owed everything to the influx of rural surpluses, who invariably proved tragically infection-prone, and to long-distance migrants who brought new diseases with them.

Egypt was one centre. The Old Testament chronicles the

epidemics the Lord hurled down upon the kingdom of the Pharaohs; and their melancholy toll upon Greece was also recorded. Reputedly beginning in Africa, a dreadful pestilence hit Greece in 430 BC, its impact on Athens being recounted by the historian Thucydides. Victims succumbed to headaches, coughing, vomiting, chest pains and convulsions; their flesh became red with blisters and ulcers; and the trouble descended into the bowels before death spared them further suffering. What was it? We do not know, but it was so catastrophic that it spelt an end to the ascendency of Athens.

Epidemics worsened with the sway of Rome. As its legions conquered the known world, deadly pathogens were given free passage around the Empire, coming home to the Eternal City itself. The Antonine plague – probably smallpox, long smouldering in Africa and Asia – slew a quarter of the inhabitants in stricken areas between AD 165 and 180, some five million in all.

Wherever it pounced on a virgin population, measles too proved lethal. In his *Observations Made During the Epidemic of Measles on the Faroe Islands in the Year 1846*, a well-documented recent instance of one such strike, Peter Panum reported how the disease had affected no fewer than 6,100 out of 7,864 inhabitants on that remote Atlantic island which had been completely free of the disorder for the previous sixty-five years.

From being mass killers, measles, chickenpox and the like have, of course, abated into routine and generally mild diseases of childhood. Time was when a virgin region would be buffeted by lethal epidemics of these diseases which killed or immunized so many that the pathogens themselves died out for lack of hosts – a form of counter-productive microbial overkill probably evident in the Athenian plague. But eventually such centres became large enough to house sufficient non-immune individuals for them to host the diseases permanently – for this an annual case-total of something in the range of 5,000–40,000 may be necessary. In such circumstances, conditions such as measles attenuated into illnesses of childhood which, because of mother-conveyed immunities, generally affect the young less severely and confer resistance against future attacks. As diseases which initially had been murderously epidemic and turned endemic, expanding populations accommodated and surmounted them; but the diseases gained a permanent foothold, becoming, if less lethal, lastingly debilitating.

Populations moreover remained exposed to other dire infections, especially certain ones with insect vectors, against which humans remained immunologically defenceless, because they were primarily afflictions not of humans at all but of animals. One such is bubonic plague, basically a rodent disease. The plague bacillus strikes humans only when infected fleas, having killed off the whole of the preferred rat population in an epizootic, are forced to turn on human victims, with devastating effects. When the flea bites its host, the bacillus enters the bloodstream. Filtered through the nearest lymph node, it produces the characteristic swelling ('bubo') in the neck, groin or armpit, killing within days about two-thirds of those infected.

The first documented bubonic plague outbreak occurred, predictably enough, in the Roman Empire. The plague of Justinian originated in Egypt in AD 540; two years later it blitzed Constantinople and massacred a quarter of the population of the eastern Mediterranean. It was a subsequent plague cycle, however, which had the most devastating impact. Towards 1300 the Black Death began rampaging through Asia before sweeping westwards through the Middle East to North Africa and Europe, replacing leprosy as the scourge of God. Between 1346 and 1350 it felled perhaps 20 million, around a quarter of Europe's population – the largest tally of fatalities caused by a single epidemic in European history. Plague dug in, and it

2. *A physician dressed in protective plague costume.*
Line engraving, after Manget.

fuelled those ghastly bogeys which haunted the late-medieval Gothic imagination – fearsome visions of Hell and the Devil, the *danse macabre*, the Horsemen of the Apocalypse, the Grim Reaper – and provoked heretic-hunting and the witch-craze among miserable sinners convinced they must placate God.

Trade, war and conquest have always exported disease. The most cataclysmic event ever for human health was Columbus's landfall on Hispaniola (today's Dominican Republic and Haiti). Contact was made in 1492 between two human populations, Old and New World, isolated from one another for thousands of years, and the biological consequences were truly catastrophic. Indigenous New World peoples formed a vulnerable virgin population, entirely without resistance to the diseases imported by the Spanish *conquistadores*.

The first New World epidemic, which struck Hispaniola in 1493, may well have been swine influenza, carried by pigs aboard Columbus's ships. Others followed. Reaching the Caribbean in 1518, smallpox killed one-third to one-half of the Arawaks on Hispaniola and spread to Puerto Rico and Cuba. It also went with Cortés to Mexico in 1521. The Spanish adventurer attacked the main Aztec city, Tenochtitlán (modern Mexico City), with just 300 Europeans and some allies. When the city fell three months later, half of its 300,000 inhabitants had died, including the Aztec leader, Montezuma, mainly from the disease. The same happened when Pizarro took on the Incas ten years later: smallpox ran ahead of him to Peru, and did much of his dirty work for him.

And this was only the beginning of a prolonged germ onslaught unleashed against the Amerindians. Waves of measles, influenza and typhus followed, all bringing devastating mortalities. Though the mainland populations of Mexico and the Andes recovered, in the Caribbean and in parts of Brazil decline neared extinction point, and the conquering Spaniards and

3. A monster representing the influenza virus.
Pen and drink drawing, E. Noble, 1918.

Portuguese were soon driven to importing slaves from Africa to meet the labour shortages created by catastrophic mortality. This trade in turn brought malaria and yellow fever, creating yet further disasters. Guns and germs together enabled tiny European forces to conquer half a continent.

In this Columbian exchange, Columbus conceivably brought back one dire disease from the Americas: syphilis. Europe's first attack broke out in 1493–4 at the siege of Naples during a Franco–Spanish conflict being waged over Italy. Soon a terrible epidemic was raging. It began with genital sores, progressing to a rash, ulceration and revolting abscesses, eating into bones and nose, lips and genitals, and often proving deadly.

The fact that some of the Spanish soldiers had accompanied Columbus suggested an American origin for the 'great pox' (to distinguish it from smallpox). As is characteristic with a new disease, for a couple of decades it spread like wildfire. 'In recent times,' reflected one sufferer, Joseph Gruenpeck,

> I have seen scourges, horrible sicknesses and many infirmities affect mankind from all corners of the earth. Amongst them has crept in, from the western shores of Gaul, a disease which is so cruel, so distressing, so appalling that until now nothing so horrifying, nothing more terrible or disgusting, has ever been known on this earth.

One of several diseases caused by the *Treponema* group of spirochetes (a corkscrew-shaped bacterium), syphilis was typical of the new plagues of an era of disturbance and migrations, being spread by international warfare, surging population and the movements of soldiers and refugees.

In a later age typhus replaced syphilis as the great wartime killer, a classic disease of dirty camps and ill-kempt soldiers. In alliance with 'General Winter', it turned Napoleon's Russian invasion into a disaster. The French crossed into Russia in June 1812, and the emperor reached Moscow in September to find

4. *Young suitor kneeling before death disguised as a young girl. A satire on syphilis.*

the city abandoned. During the next five weeks, the *grande armée* suffered a devastating typhus epidemic. Of its 600,000 men, few got back, and typhus was a prime cause. Even then it was becoming one of the great 'filth diseases' of the shock towns of the Industrial Revolution.

Cholera, however, was the new disease of the nineteenth century. Endemic to the Indian subcontinent, cholera had never gone global. Beginning in 1816, the first pandemic raged in Asia, moved west and threatened to enter Europe, but receded. The second began in 1829. It spread through Asia, broke into Egypt and North Africa, entered Russia, tracked across Europe, and familiarized a ghastly way to die. Acute nausea led to violent vomiting and diarrhoea, with stools turning to a grey liquid described as 'rice water' until nothing emerged but liquid and fragments of gut. Extreme cramps followed, with an insatiable desire for water, followed by a sinking stage. Dehydrated and nearing death, the patient displayed the classic cholera physiognomy: puckered blue lips in a shrivelled hollow face.

There was no agreement about its cause; many treatments were touted; nothing worked. London was hit in 1832 with 7,000 dying, and so was Paris. Cholera reached North America in the same year, first attacking New York and the eastern seaboard, by 1834 crossing to the Pacific and spreading south to Latin America.

The third pandemic began in 1852, and 1854 proved a dreadful year. Between 1847 and 1861, two and a half million Russians contracted the disease and over a million died. The fourth pandemic started in 1863 and lasted until 1875, and the fifth brought devastation to Hamburg in 1892 (a faulty piped-water system made things worse). By that time, however, cholera could be controlled through public-health measures, especially after Robert Koch's isolation of the bacillus in 1884. As a consequence, the sixth pandemic (1899 to 1926) barely

5. A young Venetian woman, aged twenty-three, depicted before and after contracting cholera.

affected western Europe. Recent years have seen returns of cholera outside Asia, notably in Latin America.

If agriculture proved a mixed blessing – it enabled larger numbers to survive, albeit with compromised vitality – the Industrial Revolution brought similar trade-offs. While bringing population growth and greater prosperity (if inequality as well), industrialization also spread insanitary living conditions, occupational diseases (such as the lung diseases of miners and potters) and new urban conditions such as rickets.

And alongside the old diseases of poverty there emerged diseases of affluence. Cancer, obesity, coronary heart disease, hypertension, diabetes, emphysema and many chronic and degenerative conditions mushroomed among wealthy, ageing nations, and they are now beginning a rampage through the Third World as Western lifestyles are exported, with cigarettes, alcohol, fatty diets, junk food and narcotics taking their toll in Asia, Africa and Latin America.

Though cholera and other killers receded, the twentieth century brought new ones. The 'Spanish flu', which swept the globe in the aftermath of the Great War, was the worst pandemic ever, slaughtering perhaps 60 million people worldwide in less than two years. (Its precise cause remains unknown, prompting fears that deadly flu might return.) And new diseases still make their appearance: AIDS, Ebola, Lassa and Marburg fevers, for example. Originating in sub-Saharan Africa, AIDS, transmitted through sexual fluids and blood, first came to medical attention in 1981, when it was found that homosexual men in America were dying from rare conditions associated with immune system breakdown. A panic period marked by victim-blaming ('gay plague'), political buck-passing and intensive medical research was followed in 1984 by the discovery of the human immunodeficiency virus (HIV), today almost universally held to

6. An over-indulgent man inflicted with the gout. The pain is represented by a demon burning his foot. Engraving, G. Cruikshank, 1818.

be responsible for the condition. Hopes for a vaccine or a cure have been frustrated, however, partly because the virus mutates so fast: drug treatments so far remain only palliative. Moreover, because HIV breaks down the immune system, sufferers are liable to opportunistic infections, helping such diseases as tuberculosis, recently thought eradicated, to make a comeback. Exceedingly dangerous as a result of being long asymptomatic, AIDS remains out of control, and is at its most devastating in those nations of sub-Saharan Africa which are poorest and have the scantiest medical resources.

In 1969 the US Surgeon General told the American nation that the book of infectious disease was now closed: the antimicrobial war had been won. The folly of that view is a measure of the myopic medical optimism so prevalent a generation ago. Today's mood is much more sober. From an evolutionary perspective, man's global fight against disease seems more like a holding operation in a war without end.

Until recent times life was everywhere lived under the empire of disease. Up to half of all babies born did not survive infancy, childhood and adolescence were highly vulnerable periods and tragically vast numbers of mothers died in childbirth. 'The world is a great hospital' was a proverbial expression. Such experiences coloured the Christian vision of the world as a vale of tears: man *must* be sinful – why otherwise could there be so much suffering?

People, the poor especially, had to harden themselves to sickness, pain, disability and premature ageing. Stoicism became second nature, but not fatalism: our forebears tried to keep themselves well, and to care for themselves and their families when sick. And those who could afford it, sometimes turned to professional healers.

CHAPTER TWO

Doctors

> His neckerchief and shirt-frill were ever of the whitest; his clothes were of the blackest and sleekest; his gold watch chain of the heaviest and his seals of the largest. His boots, which were always of the brightest, creaked as he walked . . . and he had a peculiar way of smacking his lips and saying 'Ah' at intervals while patients detailed their symptoms, which inspired great confidence.
>
> Charles Dickens (1812–70), *Martin Chuzzlewit*
> (description of Dr Jobling, general practitioner)

Emerging in a disease-riddled environment, civilization sought forms of propitiation and relief. People have always tried to protect themselves and their families – that is integral to self-preservation and parenting. But from early times, healing also became the craft of diviners and witch-doctors, fighting off the disorders raining down from above and offering remedies. Ancient cave paintings, some 17,000 years old, depict men masked in animal heads, performing ritual dances; these may be our oldest images of medicine-men. With the evolution of more complex settled societies, herbalists, birth-attendants, bone-setters and healer-priests followed.

Distinctive among indigenous healers is the shaman, common in Siberia and the New World, with his repertoire of magic and rituals against disease. Deploying fetishes, amulets to protect against black magic and talismans for good luck, shamans combined the roles of healer, sorcerer, seer, teacher and priest, and claimed spiritual powers to heal the sick, combat sorcery and

7. An African medicine man or shaman using symbols and small animals to eject a demon (disease). Wood engraving, after J. Leech.

ensure fertility. Shamans and similar folk healers are now credited by anthropologists with valuable skills, both medical and social.

With the rise of settled civilizations, healing practices grew more elaborate and were written down. In ancient Mesopotamia (Iraq), an official medical system emerged based on a diagnostic framework which drew on omens and divination techniques, including hepatoscopy, the inspection of the livers of sacrificed animals. Treatments combined religious rites and empirical treatments. Under a head physician, three types of healers practised: a seer (*bârû*), who was expert in divination; a priest (*âshipu*), who carried out exorcisms and incantations; and a physician (*âsû*), who employed drugs and performed surgery and bandaging.

As in Mesopotamia so in the Egypt of the Pharaohs (third millennium BC onwards), the *swnu* (physician) formed one of a three-fold public division of healers, the others being priests and sorcerers. One such physician was Iri, Keeper of the Royal Rectum, the Pharaoh's enema expert; another was Peseshet, head female physician – confirmation of the presence, as in the Middle East, of women healers. Most famous was Imhotep, chief vizier to Pharaoh Zozer (2980–2900 BC), renowned as a physician, astrologer, priest, sage and pyramid designer. His 'sayings' were later written down, and within a few generations he was being deified. As surviving papyri show, Egyptian medicine combined religious beliefs and magical techniques with an impressive array of practical drug treatments and surgical skills.

Among the Greeks, various gods and heroes were identified with health and disease, the chief being Asklepios (Aesculapius in Latin), a figure similar to Imhotep. Homer depicted him as a tribal wound-healer, though he became widely hailed as a son of Apollo, the god of healing. Elevated into the patron saint of

8. *Figure of Asklepios. Etching, N. Dorigny.*

medicine, the bearded Asklepios was portrayed with a staff and snake – the origin of the modern caduceus sign, with its two snakes intertwined, like a double helix, on a winged staff. He was often shown accompanied by his daughters, Hygeia (health) and Panacea (cure-all), and his sons supposedly became the first physicians (Asklepiads). The cult of Asklepios spread, and by 200 BC every Greek city-state (*polis*) had its temple to the god, the best known being those on the island of Cos, Hippocrates' reputed birthplace, and at Epidaurus, thirty miles from Athens. As in Egypt, sick pilgrims would sleep overnight in special incubation chambers where, before an image of Asklepios, they hoped to receive a healing vision in a dream.

In a break with these sacred practices, the first appearance in the West of an essentially secular medicine came with the Hippocratic doctors who emerged in the Greek-speaking world in the fifth century BC. Decrying traditional and religious healers, they developed an elitist ideal of professional identity. Elevating themselves above root-gatherers, diviners and others whom they dismissed as ignoramuses and quacks, the Hippocratics promoted *natural* theories of health and sickness (grounded upon superior *natural* knowledge), and *natural* modes of healing. No longer pretending to be an intercessor with the gods, the true doctor would be the wise and trusty bedside friend of the sick.

Legend has it that Hippocrates (*c.* 460–377 BC) was born on the island of Cos, and that he was a fount of medical wisdom and an honourable man. The sixty or so works which comprise the so-called Hippocratic *corpus* were penned by him only in the sense that the *Iliad* is credited to Homer or the Pentateuch to Moses. Internal discrepancies show that they derive from a variety of hands over a period of time.

Rather as in Indian Ayurvedic medicine, the *corpus* broadly explained health and illness in terms of the humours. The body was subject to rhythms of development and change which were determined by key fluids (humours) constrained within the skin envelope; health or illness resulted from their shifting balance. These crucial vitality-sustaining juices were blood, choler (or yellow bile), phlegm and black bile. The four served different life-sustaining ends. Blood was the source of vitality. Choler or bile was the gastric juice, indispensable for digestion. Phlegm, a broad category comprehending all colourless secretions, was a lubricant and coolant. Also visible in sweat and tears, it was most evident when in excess – at times of cold and fever. The fourth fluid, black bile, or melancholy, was more problematic. A dark liquid almost never found pure, it was reckoned responsible for darkening other fluids, as when blood, skin or stools turned brackish.

Between them, the four major fluids accounted for the visible and tangible phenomena of physical existence: temperature, colour and skin texture. Blood made the body hot and wet, choler hot and dry, phlegm cold and wet, and black bile produced cold and dry sensations. Parallels were drawn with the four elements discerned by Greek science in the universe at large. Being hot and agitated, blood was like air; choler was like fire (hot and dry); phlegm suggested water, and black bile (melancholy) resembled earth (cold and dry). Such analogies further pointed to and meshed with other facets of the natural world, including astrological influences and seasonal variations. Cold and wet, winter thus had affinities with phlegm; it was the time people caught chills.

Each fluid also had its distinctive colour – blood being red, choler yellow, phlegm pale and melancholy dark. These hues were responsible for body coloration, giving vital clues as to why different peoples were distinctively white, black, red or yellow, and why certain individuals were paler, swarthier or ruddier than others.

Humoral balance was also responsible for bodily shape and physique: phlegmatic people tended to be fat, for example, the choleric ones thin. It further explained the temperaments, or what, in later centuries, would be called personality and psychological dispositions. Thus someone generously endowed with blood would present a florid complexion and have a sanguine temperament, being lively, energetic and robust, though perhaps given to impulsive hot-bloodedness. Someone cursed with surplus choler or bile might be choleric or acrimonious, quick to anger and marked by an acid tongue. Likewise with phlegm (pale, and phlegmatic, lazy, inert and cool in character) and black bile (one with swarthy looks and a saturnine disposition – that is, sardonic, suspicious, prone to look on the dark side). There was, in short, infinite and flexible explanatory potential in such rich holistic linkages of physiology, disposition and presence, not least because convincing links were suggested between inner constitutional states (temper) and outer physical manifestations (complexion or, in the sick, disease symptoms): such beliefs were not just plausible but quite indispensable so long as science and medicine had scant direct knowledge of what went on beneath the skin.

Humoral thinking also had ready explanations when people fell sick. All was well when the vital fluids co-existed in a proper balance. Illness resulted, however, when one of them built up (became plethoric) or diminished. If, perhaps through faulty diet, the body made too much blood, sanguineous disorders followed as one grew overheated and feverish. One might, by consequence, have a seizure, an apoplectic fit, or grow maniacal. Deficiency of blood or poor blood quality, by contrast, meant reduced vitality, while blood loss due to wounds would lead to fainting, coma and even death.

Fortunately, held the Hippocratic authors, these imbalances were capable of prevention or correction through sensible

9. The Four Humours, fifteenth century.

10. Four heads of men who represent each of the four temperaments. Engraving, W. Johnson, early nineteenth century.

lifestyle (regimen), or by medical or surgical means. The person whose liver concocted a surfeit of blood or whose blood was thought polluted with toxins should undergo blood-letting. Change of diet could help too. Detailed recommendations were spelt out for regulating exercise, and diet (collectively known as 'dietetics'): prevention was better than cure.

The appeal of the humoralism which dominated classical medicine and formed its heritage lay in its comprehensive explanatory scheme, which drew upon bold archetypal contrasts (hot/cold, wet/dry, etc.) and embraced the natural and the human, the physical and the mental, the healthy and the pathological. While reassuringly intelligible to the layman, it was a supple tool in the hands of the watchful bedside physician and open to further theoretical elaboration.

Hippocratic doctors made no pretence to miracle cures, but they did pledge above all to do no harm (*primum non nocere*) and presented themselves as faithful friends to the sick. This humane disposition demonstrated the physician's devotion to his art rather than fame or fortune, and consoled anxious patients. Ethical concerns about medical conduct were addressed in the Hippocratic Oath.

The Oath

I swear by Apollo the healer, by Aesculapius, by Health and all the powers of healing, and call to witness all the gods and goddesses that I may keep this Oath and Promise to the best of my ability and judgment.

I will pay the same respect to my master in the Science as to my parents and share my life with him and pay all my debts to him. I will regard his sons as my brothers and teach them the Science, if they desire to learn it, without fee or contract. I will hand on precepts, lectures and all other learning to my sons, to those of my master and to those pupils duly apprenticed and sworn, and to none other.

I will use my power to help the sick to the best of my ability and judgment; I will abstain from harming or wronging any man by it.

I will not give a fatal draught to anyone if I am asked, nor will I suggest any such thing. Neither will I give a woman means to procure an abortion.

I will be chaste and religious in my life and in my practice.

I will not cut, even for the stone, but I will leave such procedures to the practitioners of that craft.

Whenever I go into a house, I will go to help the sick and never with the intention of doing harm or injury. I will not abuse my position to indulge in sexual contacts with the bodies of women or of men, whether they be freemen or slaves.

Whatever I see or hear, professionally or privately, which ought not to be divulged, I will keep secret and tell no one.

If, therefore, I observe this Oath and do not violate it, may I prosper both in my life and in my profession, earning good repute among all men for all time. If I transgress and forswear this Oath, may my lot be otherwise.

As is clear, the Oath was intended to protect doctors, through a guild-like closed shop, no less than to safeguard patients. With its assumption of benevolent sagacity, it underwrote the medical profession's lasting paternalism.

For all its latter sacred status, little is known about the Oath's origins or early use. It obviously foreshadows, however, the paradigm of a profession (one professing an oath) as an ethically self-regulating discipline among those sharing specialized knowledge and committed to a service ideal. As it makes clear, Hippocratic medicine was a male monopoly, although physicians expected to cooperate with midwives and nurses.

Hippocratic medicine had its weaknesses. It knew little anatomy or physiology, since human dissection would have countered Greek respect for the human; and it lacked effective cures. Its strong point, however, and its lasting attraction, lay in casting sickness as a disturbance in the individual, who would then be granted personal medical attention. 'Life is short, the art long, opportunity fleeting, experience fallacious, judgment difficult', proclaims the first of the Hippocratic aphorisms, thus outlining the physician's demanding but honourable calling. This lofty ideal commands respect to this day as a paradigm for professional identity and conduct.

If Hippocrates is shadowy, Galen, the 'emperor' of medicine under the Roman Empire, is high-profile. His egotism and omniscience, and the sheer bulk of his surviving writings, ensured that his authority dominated medicine for nearly a millennium and a half.

The son of a wealthy architect, Galen (AD 129–*c.* 216) was born in Pergamon (modern Bergama, Turkey). When he was sixteen, his father, we are told, was visited in a dream by Asklepios, after which the son was piously steered towards medicine. In 162 he left for Rome, where dazzling performances of his anatomical skills spread his fame. He was soon in imperial service.

Expert in one-upmanship, Galen hid his self-importance under the cloak of the dignity of medicine, while scolding colleagues and rivals as ignorant buffoons. Philosophy, he taught, was essential to endow medicine with the theoretical basis it required. The physician should not be a mere practical healer (empiric) but must master logic (the art of thinking), physics (the science of nature) and ethics (the rule of action). The unphilosophical healer was like a botching builder: the true physician should be like an architect armed with proper blueprints.

The patient's trust, so essential to healing, could be won by a good bedside manner and by mastery of prognosis, an art demanding observation, logic and experience. Indeed, Galen prided himself on being more than a superior clinician: he was a man of science, skilled in performing dissection – not of human corpses admittedly, but of apes, sheep, pigs and goats and even an elephant's heart. He developed skeletal anatomy and an understanding of the nerves, but, human dissection being highly controversial, little internal human anatomy. Exactly as he expected, Galenic medicine proved epochal. 'I have done as much for medicine,' he boasted,

> as Trajan did for the Roman Empire when he built bridges and roads through Italy. It is I, and I alone, who have revealed the true path of medicine. It must be admitted that Hippocrates already staked out this path . . . he prepared the way, but I have made it passable.

With the Christianization of the Roman Empire, medicine and religion overlapped, coalesced and occasionally clashed. Some early Church fathers deprecated pagan medicine, and it was long a smart gibe that *ubi tre physici, due athei* (where there are three physicians, there are two atheists). Echoing the Greek cult of Asklepios, Christian healing shrines flourished and saints and martyrs were invoked for health. Each bodily organ and complaint acquired its particular saint – St Anthony for erysipelas, St Vitus for chorea, and so forth. Supplanting Asklepios, St Damian and St Cosmas became the patron saints of medicine at large.

In the so-called Dark Ages, healing became the preserve of monks and clerics, the only learned men left in the West. The flame of classical medicine was meanwhile kept alive in the far more advanced Islamic world, where a succession of distinguished scholar-physicians, active in what are modern Syria, Iraq, Iran, Egypt and Spain, studied, further systematized, and extended the work of Galen.

From the twelfth century, however, with the founding of universities and the recovery and retranslation of learned medicine from Islamic sources, professional medicine itself recovered, initially at Salerno in southern Italy. Education was based on set texts, formalized by the new Aristotelian scholasticism. After seven years spent attending lectures and engaging in disputations and oral examinations, a student could graduate as a qualified physician. The goal of a formal scholastic medical education lay in the acquisition of rational knowledge (*scientia*) within a philosophical framework: the learned physician who knew the reasons for things would not be mistaken for a mere 'empirical' healer or a quack. There were few such Galenic paragons, though: most medieval practitioners picked up their skills by apprenticeship and experience.

Through the Middle Ages into the Renaissance and long beyond, the ideal physician was prized as a man – the profession remained a male monopoly – who had undergone a prolonged university education to render him expert in the liberal arts and sciences; he would be upright, trustworthy and God-fearing, grave, sober and devoted to learning not lucre. '*Hippocrates*,' pronounced James Primrose in 1651, engaging in typical ancestor-worship, 'saies that a physician which is a Philosopher, is God-like.' 'Physicians, like beer,' opined Thomas Fuller, 'are best when they are old.'

To set off this saintly figure of the ideal physician as high-minded, dignified and austere, his antitheses were vilified – the money-grubbing pretender; the swindling quack (a 'turdy-facy, nasty-paty, lousy-fartical rogue', according to Ben Jonson); the tipsy nurse; the greasy, gossiping midwife. The traditional surgeon was often caricatured as a man of the flesh – bold and beefy, handy with the knife and saw, little better than a butcher and no more learned than the barber, with whose trade he frequently doubled. The superior physician plumed himself as being marked out by mind not muscle, brains not brawn.

11. Surgeons participating in the amputation of a man's lower leg. Aquatint, Thomas Rowlandson, 1793.

Across Europe an image of a sound consultation as practised by such a man remained entrenched right down to the nineteenth century. By cross-questioning, the physician would ascertain the symptoms (taking the patient's history), establish the nature of the disease, frame a diagnosis and formulate a regimen. This would probably include prescribing herbal drugs, to be compounded by the apothecary – alongside the surgeon, another lesser light of the profession. Before the introduction of systematic physical examinations and diagnostic tests, the physician's job was not hands-on: what counted were book-learning, experience, memory, judgement and a good bedside manner. The deeply traditional veneer of medicine made it comforting – or, to satirists, antiquated and ridiculous.

As the number of doctors rose, medicine became organized, first in urban Italy, where guilds emerged and assumed responsibilities for apprenticeship, examination of candidates, oversight of pharmacists and supervision of drugs. Medical organization took various forms. As early as 1236 Florentine physicians and pharmacists banded into a single guild, recognized as one of the city's seven major crafts. In southern Europe no great divide opened between surgeon and physician. Elsewhere a social and professional gulf widened, for beyond Italy surgery was excluded from the academic curriculum. In northern Europe it was tied to barbering and regarded by physicians as rather infra dig.

In London the Fellowship of Surgeons came into being in 1368–9, and a Company of Barbers was chartered in 1376. The founding of the College of Physicians of London in 1518 (it became *Royal* at the restoration of Charles II) authorized the physicians to regulate metropolitan practice. In time, all such medical colleges and corporations received a bad press as monopolistic oligarchies protecting the privileged against the interests of patients and lesser practitioners alike.

*

Partly to assuage the terrors of diseases which it could all too rarely cure, nineteenth-century primary care stuck to reassuringly familiar public practices. The paying private patient would summon a doctor of choice (traditionally by sending a servant, but after 1900 perhaps by telephone), who would then pay a house call – on horseback, by pony-and-trap or, in the twentieth century, increasingly by motor-car. Relations between patients and family doctors were personal and governed by the strict protocols of gentlemanly behaviour; social graces counted.

There were grumbles on both sides – notably about uppity doctors, and unpaid bills – but the profession had a stake in nurturing family care, even in cossetting such tiresome valetudinarians as Mr Woodhouse in Jane Austen's *Emma* who comprised the 'worried well'. Cynics insinuated that physicians sowed habits of sickness among their better-off patients, in particular the weaker sex, trading in fancy diagnostic jargon, favourite prescriptions, the minutiae of diet and lifestyle, and all the other rituals of a profession that found it paid to be obsequious to the carriage trade. A *Punch* cartoon of 1884 featured a conversation:

FIRST LADY: What sort of a doctor is he?

SECOND LADY: Oh, well, I don't know very much about his ability; but he's got a very good *bedside manner*!

All such palaver veiled the fact that, right through into the twentieth century, the 'disease empire' discussed in the previous chapter called the shots. Families were assailed by a battery of infections and fevers which might well prove lethal; gastrointestinal and dysenteric troubles, diphtheria, chickenpox, scarlet fever and rubella claimed hordes of infants, while measles, tuberculosis, syphilis, meningitis and childbed fever were part of the common round of the average physician.

In this situation, the old-style doctor had a choice between the

ANNALS OF A WINTER HEALTH RESORT.

Lady Visitor. "OH, THAT 'S YOUR DOCTOR, IS IT ? WHAT SORT OF A DOCTOR, IS HE ?"

Lady Resident. "OH, WELL, I DON'T KNOW MUCH ABOUT HIS ABILITY ; BUT HE 'S GOT A VERY GOOD BEDSIDE MANNER !"

12. *'Annals of a Winter Health Resort',* Punch *cartoon, 1850.*

conservative Hippocratic options (waiting and watching, bed-rest, tonics, care, soothing words, calm and hope), or 'heroic' possibilities, including violent purges, drastic blood-letting (Galen's preference), or some pet nostrum of his own. Often his decision was made for him: crusty patients had strong opinions about the right treatment for 'their' illnesses, and he who paid the piper called the tune.

Primary care's options were in any case limited, since, before the twentieth century, the pharmacopoeia resembled a box of blanks. Of the thousands of medicaments in official use, few were truly effective: among these were quinine for malaria, opium as an analgesic, colchicum for gout, digitalis to stimulate the heart, amyl nitrate to dilate the arteries in angina and, introduced in 1896, the versatile aspirin. Iron was ladled out as a tonic, as were senna and other herbal preparations as purgatives. True cures remained elusive, however, and doctors knew their prescriptions were largely eyewash. This dismal situation was somewhat allayed by the fact that churchgoing folk did not expect the family doctor to perform miracles and, living in a vale of tears, they were inured to a constant round of funerals. In a famous Victorian painting by Luke Fildes a physician sits by the bedside of a dying child, unable to do anything but show care and compassion: the tone of the portrait is not accusatory but sympathetic.

Teeth tightly clenched, elite medical professors could espouse a grim therapeutic nihilism: medicine could understand the diseases from which people would die but not stop them from dying. But family doctors inevitably felt pressed to do something. That explains the growing recourse to the strong sedatives, analgesics and narcotics newly marketed by the nineteenth-century pharmaceutical companies. Thanks to the synthesis of morphine in 1806 and the invention of the hypodermic syringe in 1853, it became easy to give fast fixes of

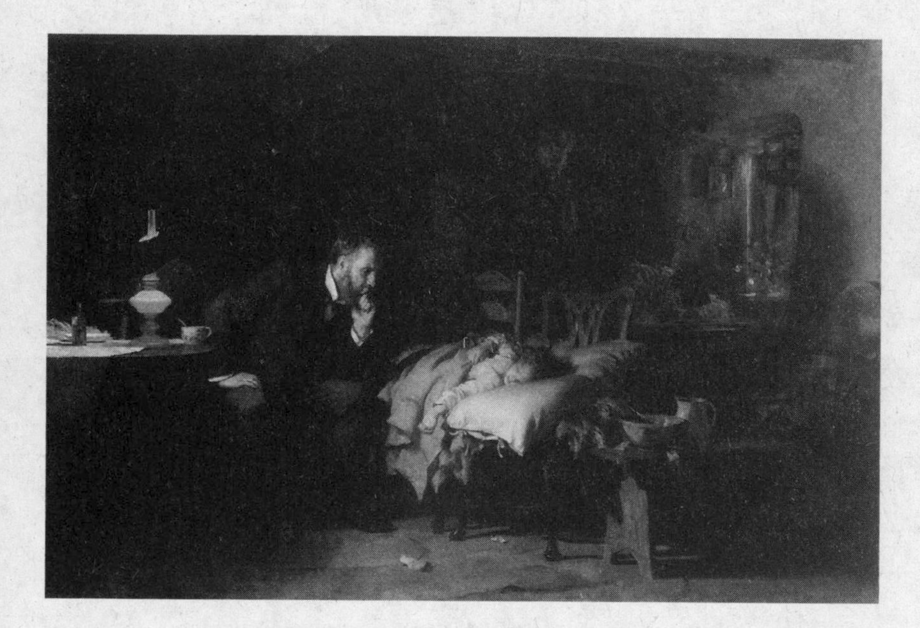

13. The Doctor. *Luke Fildes, 1891.*

strong opiates – eventually even the newly developed heroin, introduced by Bayer in 1898. In 1869 chloral hydrate came into use as a sleeping potion; barbitone (Veronal) appeared in 1903 and phenobarbitone in 1912. Painkilling at least became possible, at the cost in many cases of addiction.

If his ability to heal the sick remained patchy, the GP consolidated his position by developing his skills. In his charming autobiography *The Horse and Buggy Doctor* Arthur Hertzler, a small town physician born in Iowa in 1870, noted in 1938 the changes which had taken place in his own lifetime. This was bedside medicine, old style:

> The usual procedure for a doctor when he reached the patient's house was to greet the grandmother and aunts effusively and pat all the kids on the head before approaching the bedside. He greeted the patient with a grave look and a pleasant joke. He felt the pulse and inspected the tongue, and asked where it hurt. This done, he was ready to deliver an opinion and prescribe his pet remedy.

Fresh back from advanced Berlin, young Doc Hertzler was minded to make his own practice more scientific, through giving rigorous and systematic physical examinations. This would boost his rating if not his cure rate: 'I had ideas of my own,' he declared. His new attempts at

> physical examination impressed my patients and annoyed my competitors, which, of course, I accepted as a two-time strike. Word went out that the young doctor 'ain't very civil but he is thorough'. Only yesterday one of my old patients recalled that when I came to see her young son I 'stripped him all off and examined him all over'. Members of that family have been my patients for the intervening forty years, so impressed were they.

New-fangled apparatus steadily contributed to the coming ideal of the thorough physical examination and, later, the check-up. First the stethoscope, invented in 1816, and subsequently devices such as the ophthalmoscope and the laryngoscope (mid-Victorian) imparted a new meticulousness (and mystique) to the business of diagnosis. From the 1860s compact thermometers were available to measure body temperature; fever charts permitted plotting of the temperature patterns typical of specific diseases; and sphygmomanometers allowed the testing of blood-pressure. The early twentieth-century GP with access to a diagnostic laboratory might also examine bodily fluids – and increasingly that meant searching for microbes, the enemy revealed by the dazzling science of bacteriology and its gospel of germs. Most patients, like Hertzler's, welcomed these extensions to the physical examination – though some resented their intrusiveness. Dr Arthur Conan Doyle, Sherlock Holmes's creator, recorded in 1881 a 'frightful horror' of a patient, who would not let him examine her chest: 'Young doctors take such liberties, you know, my dear.'

'Scientific medicine' was most keenly taken up in the USA, more eager about technology. 'Working with the microscope and making analyses of the urine, sputum, blood, and other fluids as an aid to diagnosis,' reflected a hard-bitten American physician in 1924, 'will not only bring fees and lead to valuable information regarding your patient's condition, but will also give you reputation and professional respect.' His Old World equivalents, by contrast, were more cautious. When the eminent British physician Sir James Mackenzie pronounced in 1918 that 'laboratory training *unfits* a man for his work as a physician', he was speaking for much of his profession – and probably his patients as well.

The likes of Mackenzie knew that the hallowed rituals of bedside medicine upheld the sacred personal bond between physician and patient. In the reign of Queen Victoria – or as late as the Second World War, for that matter – the best-respected general practitioners and Harley Street consultants alike were those who could impress upon patients that they were skilful, serious, attentive, trustworthy and doing their best. The Hippocratic ideal was revered, and it helped generate the 'patient-as-a-person' movement, influential after 1900 in reaction against the more scientific medicine promoted by the universities and their research laboratories. The physician, it emphasized, must see the patient as an individual. 'Never forget that it is not pneumonia, but a pneumonic man who is your patient,' declared Sir William Gull. 'The good physician treats the disease,' taught the distinguished Canadian medical humanist William Osler, 'but the great physician treats the patient.' Similar views were advanced in 1957 by the psychoanalytically inclined Hungarian-born Michael Balint, whose *The Doctor, the Patient and the Illness* extolled the apostolic function of the physician and urged that primary-care physicians should in effect become psychotherapists.

Amidst these tensions – should healing remain an art or become more scientific? – the twentieth century brought a widespread shift of gravity in the profession from general practitioner to specialist. Here a rift opened up between the UK and the USA. In Britain, primary care was to remain firmly in the hands of generalist family doctors. This was because panel practice under the National Insurance Act of 1911, later reinforced by the National Health Service (1948) – see Chapter 8 – made GPs the linchpins of a publicly funded medical system. Denied the right to attend patients in hospital, they were cut off from surgery and science and all that they implied in terms of innovations and superior professional identities. Yet GPs remained the dispensers of family care and became gatekeepers to the hospital and the specialist. On the eve of the Second World War there were in Britain some 2,800 full-time consultants but seven times as many general practitioners. As late as

2000, among the 100,000 UK physicians, a third were GPs.

In the USA, by contrast, general practice inexorably lost out to specialism. In a competitive market milieu, the scientifically advanced paediatrician, cardiologist or oncologist gained an edge. By 1942 fewer than half of all American doctors were GPs, and by 1999, of the 800,000 physicians in the USA – a staggering total in itself! – fewer than one in ten was in family practice; GPs had gone the way of the horse-and-buggy doctor.

The role of doctors, and public expectations of them, changed during the twentieth century. The old acute infectious diseases were dwindling, and in any case could be cured from the 1930s by sulpha drugs or, from the 1940s, by antibiotics. Yet, partly because of greater longevity, further chronic and abnormal conditions were coming to light, and the population seemed to be feeling worse. Self-reported illnesses rose by 150 per cent from 1930 to 1980. The average American visited the doctor 2.9 times a year in 1930; by 2000 this had doubled. Why? Though overall healthier, individuals grew more sensitive to symptoms and more inclined, or trained, to seek help for ailments their grandparents would have dismissed as trivial or untreatable. Patients had also, meanwhile, been encouraged to expect and demand more of their doctors. The 'doing better, feeling worse' syndrome emerged, and the public, having long held doctors in respect, grew disillusioned.

Once doctors became therapeutically far more potent, thanks to antibiotics and other magic bullets, they arguably abandoned the art of pleasing their patients. Armed with more effective weapons, they tended to forget the psychological significance and benefits of the close and trusting doctor/patient relationship patients expected. In the 1980s a British NHS doctor bluntly explained the function of prescribing pills at the close of a brief consultation: 'It's a nice way of getting rid of the patient; you scribble something out and rip the thing off the pad. The ripping off is really the "Fuck off".' Doctors can now cure as never before: the public may doubt whether they care.

At the dawn of the twenty-first century public expectations of healthiness are higher than ever, partly because of media-fuelled health awareness and scares. But confidence in the medical profession – especially after such scandals as the revelation that the British GP Harold Shipman had murdered hundreds of his patients – has been shaken. In a medical world which is increasingly bureaucratic and technology-driven, the Hippocratic personal touch seems in danger of being lost.

This helps explain the revitalization of irregular medicine from the 1960s. The eighteenth century was arguably the golden age of 'quackery' – a loaded term, for when speaking of non-orthodox medicine we should not automatically impugn the motives of the irregulars nor deny their healing gifts. Far from being cynical swindlers, many were fanatics about their techniques or nostrums – witness the Scot James Graham (1745–94), who touted long life and sexual rejuvenation, to be achieved by mud-bathing and his special electrified Celestial Bed, housed at his Temple of Health off London's Strand. From the 1780s the one medicine which would truly relieve gout – it contained colchicum – was a secret remedy: the *Eau médicinale*, marketed by a French army officer, Nicolas Husson, and derided by the medical profession.

Quacks excelled in entrepreneurship and the arts of publicity. Rose's Balsamic Elixir, its vendors claimed, would cure 'the English Frenchify'd' (i.e., venereal patients) at a stroke: 'it removes all pains in 3 or 4 doses'. Itinerants became expert market-place performers: gaudily dressed and flanked by a zany on a makeshift stage, they would draw first a crowd and then perhaps some teeth, give out gratis a few bottles of julep or cordial, sell a few dozen more, and then ride out of town. Most charlatans were small-timers, but some made big killings. From his 'pill and drop', Joshua Ward (1685–1761) won not only a fortune but royal favour.

With the rise of the consuming public, demand welled up for many sorts of healing and commercial society provided openings which nostrum-mongers, rejuvenators and cancer-curers rushed to fill. The craving for sure-fire cures produced 'toadstool millionaires' galore, eager to bestow magnetic, electrical, chemical or herbal cures on the desperate and credulous. Proprietary medicines won loyal followings. 'Lydia E. Pinkham's Vegetable Compound' was sold, from 1873, by Lydia Pinkham of Lynn, Massachusetts; 'Lily the Pink' became America's first millionairess. In England James Morison made a fortune with his Vegetable Pills, and Thomas Beecham followed with his Pills and Powders. The more the state and medical authorities tried to slight or suppress them, the greater their popularity.

The nineteenth century also brought new movements grounded in principled rejection of orthodox medicine. Such alternative healing philosophies often mirrored religious dissenting sects and socio-political radicals: artisans distrustful of princes and prelates were no more disposed to swallow the medicines of privileged Colleges. Alternative healers exposed regular medicine as a closed shop, an obscurantist racket devoted to self-aggrandizement: 'a conspiracy against the laity' was George Bernard Shaw's phrase. They also condemned modern lifestyles as unnatural. Urging a return to simplicity, they praised plain living and claimed that their philosophies of health followed Nature's wholesome ways. These doctrines won their greatest following in America: medical visionaries gravitated to the New World, while fewest restrictions were imposed on practice in the new republic. Their homeland, however, was Germany.

14. A man with vegetables sprouting from all parts of his body as a result of taking J. Morison's Vegetable Pills. Lithograph, C. J. Grant, 1831.

The great trail-blazing inspiration was homeopathy, developed by Samuel Hahnemann (1755–1833), who acquired his medical education at Leipzig, Vienna and Erlangen, and imbibed an enlightened faith in the goodness of Nature. Rejecting costly polypharmacy, Hahnemann formulated his new principles. There were, he argued, two approaches to healing: the 'allopathic' treatment by opposites which informed orthodox medicine – this was misguided; and his own 'homoeopathic' approach, whose key was that 'to cure disease, we must seek medicines that can excite similar symptoms in the healthy human body'. This became the first law of homoeopathy: *similia similibus curantur* – let like be cured by like. This law of similar was supplemented by the second law, that of infinitesimals: the smaller the dose, the more efficacious the medicine. This seeming paradox followed from Hahnemann's preoccupation with drug purity and his lifelong abhorrence for the arbitrary and destructive polypharmacy of regulars. Tiny doses of absolutely pure drugs did far more good than massive doses of adulterated ones.

Another movement which prized purity was hydropathy. This originated with the Austrian Vincent Priessnitz (1799–1851), a rural prophet who, convinced of water's powers, established a spa at Gräfenberg in Silesia. Health was the body's natural condition; sickness resulted from the introduction of foreign matter; and acute disease was the body's attempt to expel such morbid material. Water treatment would bring an acute condition to a crisis, expelling poisons from the system.

Equally hostile to orthodoxy was the first of the indigenous American healing sects, Thomsonianism. Despising 'book doctors', Samuel A. Thomson (1769–1843) developed a people's-health movement touting vegetable-based therapies. His favourite was the plant *Lobelia inflata*, whose seeds caused healthy vomiting and heavy sweating. The Thomsonian gospel

15. A man being treated to a cascade of water in the name of hydrotherapy. Lithograph, C. Jacque, Paris, 1843.

was brought to England in 1838 by 'Dr' Albert Isaiah Coffin, who soon had a keen following among self-improving artisans and Dissenters, leading to a network of Friendly Botanico-Medical societies. Medical botany appealed to the self-help mentality.

Another American group, the Grahamites, dedicated themselves to healthy living via a this-worldly salvationism. The teetotaller Sylvester Graham deemed health too precious to be left to doctors. Vegetarianism and whole-grain cereals were the thing, and the 'Graham cracker' took its bow. Sexual activity was to be limited – it inflamed the passions and wasted the seminal fluid which was the quintessence of life.

Rejecting the medical nihilism of the regulars, American alternative sects were upbeat. Nature was benign and, if only people heeded her laws, bodies would naturally be well. Such was the hopeful message of osteopathy, originating in 1874 with Dr Andrew Taylor Still, who established a college at Kirksville, Missouri. Still proclaimed the body's inherent capacity to repair itself. Somewhat similar was chiropractic, established in 1895 by Daniel David Palmer, after he restored the hearing of a man by adjusting his backbone.

This radical Protestant self-help optimism was taken to its logical extreme in Christian Science. Suffocated by the Congregationalism of her parents, Mary Baker Eddy (1821–1910) spent much of her adolescence bedridden, and regular physicians did her no good. Visited by divine revelation after reading the Bible, she undertook a self-healing, whose success led her to frame her own system: 'there is but one creation, and it is wholly spiritual'. Since all was spirit and matter a phantasm, there could be no such thing as somatic disease; sickness was not in the body but in the mind, and could be cured by mental effort and faith alone. Seventh Day Adventists for their part preached abstemiousness and vegetarianism, and proclaimed a 'gospel of health', partly based on hydropathic cures. Their Health Reform Institute at Battle Creek, Michigan, was headed by John Harvey Kellogg (1852–1943), brother of the cornflake king, himself a fan of roughage.

The Nature-worshipping and spiritual preoccupations of alternative medicine highlight the shortcomings in orthodox medicine which bred a populist, anti-elitist backlash. While people wanted to be relieved and cured, they were also seeking far more from medicine – explanations of their troubles, a sense of wholeness, a key to the problems of life, new feelings, of self-respect and control. If the tenor of orthodox medicine was pessimistic, alternative medicine instilled hope.

The triumphs of regular medicine and surgery in the first half of the twentieth century brought a decline in the appeal of irregular medicine. But as medicine itself grew more bureaucratic, scientistic and apparently as authoritarian as the state-complex, the fortunes of alternative medicine revived, and new systems of massage, herbalism and spiritualism proliferated. Counter-culture critiques of Western values were also dazzled by the healing philosophies of the East. And people liked to shop around. At the end of the twentieth century there were more registered irregular healers in Britain than GPs, while in the USA, more visits were being paid each year to providers of unconventional therapy (425 million) than to primary-care physicians (388 million).

From Greek times, orthodox medicine entrenched itself as a male monopoly. Women engaged in practical healing, nursing and midwifery, of course – extensions of their domestic and mothering roles – but until the nineteenth century they were everywhere excluded from the profession as such, not least because they were barred from university attendance. The female constitution was not designed for higher education, warned male chauvinists: dominated by her womb or ovaries, a woman's place was in the home as wife and mother.

It is no accident that the first woman doctor to qualify did so in America, since that was where licensing was laxest. A Bristol sugar-refiner's daughter, Elizabeth Blackwell graduated in 1849 top of the class from the Geneva Medical School in New York. Convinced that Nature made women better healers than men, Blackwell went on to found the New York Infirmary for Indigent Women in 1857 and to organize nurses in the Civil War.

The first woman to qualify in Britain was Elizabeth Garrett, who exploited legal loopholes to gain the diploma of the Society of Apothecaries in 1865, thereby securing enrolment on the Medical Register. Within five years she had developed an extensive private practice, established St Mary's Dispensary for Women, received a medical degree from Paris, and married the wealthy James Anderson. She was instrumental in the establishment in 1874 of the London School of Medicine for Women, and by her very respectability proved a persuasive diplomat for the claims of women to be doctors.

In due course entry rights for women were won everywhere – in Germany not till the beginning of the twentieth century – but resistance remained strong. The reforms in American medical education following the Flexner Report of 1910 resulted in the closure of some women's medical schools in the USA (for being substandard), and it was only after the Second World War that the Harvard and Yale medical schools opened their doors to female students. By 1976 20 per cent of British doctors were women – though rarely at the apex of the professional pyramid – and in 1996 for the first time more than half the intake to British medical schools was female. This may presage an end to the ingrained sexism of the profession.

CHAPTER THREE

The Body

> I profess to learn and teach anatomy not from books but from dissections; not from the tenets of Philosophers but from the fabric of Nature.
>
> William Harvey

The body is pregnant with symbolic meanings, deep, intensely charged and often highly contradictory. For orthodox Christians, for instance, being originally made in God's image, it is a temple. Yet since the Fall and expulsion from the Garden of Eden, bodies have been 'vile' and the flesh weak and corrupt. The Christian body is thus both sacred and sordid. Medical beliefs are always underpinned by cultural attitudes and values about the flesh.

From earliest times, all societies have had some tangible knowledge of the innards, not least because of the practices of animal butchering and sacrifice. The Egyptians for their part perfected the art of embalming. But dissecting human corpses to further knowledge has been far from universal as a medical practice. It was not part of Hippocratic medicine – respect for the dignity of man, that microcosm of Nature, was too powerful among the Greeks; nor was it the basis of traditional medicine in India or China.

Dissection of dead humans – and, possibly, even experimenting on living slaves – first developed in Hellenistic Alexandria: the state, and the physicians its servants, had more power there. It is associated with Herophilus (*c.* 330–260 BC) and his contemporary, Erasistratus – their writings have survived only by repute. Herophilus apparently dissected human cadavers in public; he discovered and named the prostate and the duodenum (from the Greek for twelve fingers, the length of gut he found). He also seems to have been the first to grasp that the arteries are filled not (as believed) with air but with blood. But his most striking dissecting feat was the delineation of the nerves. By demonstrating that their source lay in the brain he was able to conclude that they played the role which earlier thinkers had ascribed to the arteries – that is, transmitting motor impulses from the soul (intelligence centre) to the extremities.

Erasistratus for his part experimented on living animals and perhaps humans. His main investigations concerned the brain which, like Herophilus but contrary to Aristotle, the doyen of Greek naturalists, he regarded as the seat of intelligence. Somewhat later, Galen and his contemporaries cut up dead animals and experimented on live ones. Their assumption that humans were anatomically identical to animals led to certain errors – for instance, that the liver had five lobes and the heart three ventricles.

Human dissection was not permissible within Islam, while Christian belief in the sanctity of the body (it belonged to God not man) led the Vatican to regulate the handling of corpses. In 1482, however, Pope Sixtus IV stated that, with the proviso that the cadaver came from an executed criminal and was ultimately given a Christian burial, there was no objection to dissection. Grassroots opinion long expressed deep misgivings, however: robust hostility to dissection – hated as medical profanation – made itself felt in Britain as late as the 1832 Anatomy Act, hardly surprisingly in view of the activities of grave-robbers (and of Burke and Hare) in illegally obtaining corpses for anatomists.

The journey deep into human flesh initiated by dissection is what has made Western medicine unique. It has sustained the fruitful conviction that in ever-more-minute investigation of the flesh lies the key to health and disease, even if that has also encouraged a tendency to myopic reductionism, to miss the whole by concentrating exclusively upon the parts.

The first recorded public human dissection was conducted in Bologna around 1315 by Mondino de' Luzzi, and his *Anatomia mundini* became the standard text on the subject. A brief, practical guide, *Mondino's Anatomy* was meant to be read out during an anatomy class. It addressed the parts of the body in the order in which they would be handled in dissection, beginning with the most perishable, the abdominal cavity. Looking through Galenic spectacles, he perpetuated the old errors derived from animal dissections.

Anatomy had hitherto played little part in medical education. But from Mondino's time, learned physicians began to regard it an essential grounding. Anatomy theatres were built, and public displays of human dissection regularly staged by professors. From Bologna, the practice quickly spread throughout Italy – artists too, such as Leonardo da Vinci, took it up – though anatomy teaching with a human corpse did not become routine in England and Germany before 1550.

At once spectacle, instruction and edification, public dissection was performed in winter, so as to delay putrefaction, and the corpse would be that of an executed criminal, intended as a final symbolic punishment. Early illustrations show a physician decked out in academic robes seated on a throne, intoning from an anatomical text (probably Mondino's), while a surgeon slits the cadaver open with his scalpel, and a teaching assistant indicates the relevant features with a pointer. Such book-driven anatomy – demonstration of what was already known, within the Galenic theoretical framework – provided guidance to the

16. Vesalius teaching anatomy. Andreas Vesalius, 1543.

student, who would not have had the chance to wield the blade personally or even to see much for himself.

The turning-point came with Vesalius. Born in 1514 the son of a Brussels pharmacist, Vesalius studied in Paris, Louvain and Padua, where he took his medical degree in 1537, becoming at once a professor there. In 1543 he published his exquisitely illustrated masterwork, *De Fabrica Corporis Humani* (Concerning the Construction of the Human Body), which presented accurate descriptions and illustrations of the skeleton and muscles, the nervous system, viscera and blood vessels. Commending first-hand observation, Vesalius attacked orthodox teachings on various points, and chided Galen for relying on knowledge of animal not human bodies. Though it contained no startling discoveries, the *De Fabrica* bred a new climate of enquiry: ancient dogmas were challenged, and Vesalius's successors became committed observers, vying to outshine each other in new findings.

In 1561 Gabriele Falloppio, Vesalius's student and successor at Padua, published a volume of anatomical observations with new researches on structures in the skull, ear and female genitals. It was he who coined the term vagina, described the clitoris, and delineated the tubes leading from the ovary to the uterus. Ironically, however, he failed to grasp the function of what thereafter became known as the Fallopian tubes: it was not till two centuries later that it was recognized that eggs were formed in the ovaries, passing down those tubes to the womb. It was easier to make anatomical finds than to grasp their physiological function.

By the close of the sixteenth century, anatomy in the Vesalian manner was paying rich dividends. Bartolommeo Eustachio discovered what became known as the Eustachian tube (from the throat to the middle ear) and the Eustachian valve of the heart. In 1603 Falloppio's successor at Padua, Girolamo

17. *Woodcut portrait of Vesalius. Andreas Vesalius, 1543.*

Fabrizio (Fabricius ab Aquapendente), identified the valves of the veins, a discovery which was to prove critical to William Harvey. Slightly later, Gaspare Aselli, also of Padua, drew attention to the lacteal vessels, stimulating later studies of the stomach and digestion. The knife was thus uncovering a new world of the bodily organs, though improved mapping of structures outran a correct understanding of functions: post-Vesalian anatomy still largely thought in terms of Galenic physiology.

Nevertheless, anatomy was establishing itself as the cutting-edge of medical science, and in due course the familiarity which followed from dissection drove investigators to rethink the body and its disorders – indeed, the very nature of disease. Traditional humoral theories had viewed health and disease in terms of systemic fluid balance. This model was gradually supplanted by a new concern with local anatomical structures and mechanisms – the 'solids'. The 'black box' of the body was being exposed to the medical gaze.

From earliest times blood had been treasured as the liquid of life: it was recognized as the body's nourishment or, when disordered, the source of inflammation and fever. Here, as ever, Galen was the authority. The veins that carried blood, he held, originated in the liver, while the arteries stemmed from the heart. Blood was 'concocted' (literally cooked) in the liver; it then washed outwards, like water irrigating a field, via the veins into the parts of the body, where it carried nourishment and was 'consumed' (used up). The portion of the blood proceeding from the liver to the right side (ventricle) of the heart branched into two. One small stream passed via the pulmonary artery to the lungs to nourish them, the other traversed the heart through interseptal pores into the left ventricle, where it mingled with air, became heated and proceeded to the periphery.

This model held sway for nearly a millennium and a half.

After 1500, however, as part of the new Renaissance spirit of enquiry, the master's teachings became questioned. Michael Servetus, a Spanish theologian and physician, conjectured a 'lesser circulation' which passed through the lungs: Galen's authority notwithstanding, blood could *not* flow through the septum (wall) of the heart – it was quite solid! – it must find its way from the right to the left side of the heart *across the lungs*. In 1559 Servetus's proposed 'pulmonary' circulation was given sound empirical support by the Italian anatomist, Realdo Colombo.

A yeoman's son born in Kent, William Harvey studied medicine at Caius College, Cambridge. Graduating in 1597, he pursued further studies under Fabrizio in Padua. Five years later, he set up in London as a physician, being elected Fellow of the Royal College of Physicians in 1607; and two years later he was appointed physician to St Bartholomew's Hospital.

While studying in Italy, Harvey began his investigations into the operations of the heart, and as early as 1603 he felt emboldened to assert that 'the movement of the blood occurs constantly in a circular manner and is the result of the beating of the heart'. Lectures delivered in London in 1616 show that he had confirmed Colombo's work on the pulmonary transit. In them he concluded that the heart worked as a muscle, with the ventricles expelling blood in systolic contractions rather than, as taught, sucking it in during diastole (relaxation). The arteries pulsated because of the shock-wave from the beating heart, not of their own intrinsic 'pulsative virtue'. The fruits of these new ideas were finally published in 1628 under the title *Exercitatio Anatomica de Motu Cordis et Sanguinis* (An Anatomical Disquisition Concerning the Motion of the Heart and the Blood), a work justly renowned as one of the classics of medical investigation.

Harvey opened by pointing out Galen's flaws. Discussing the

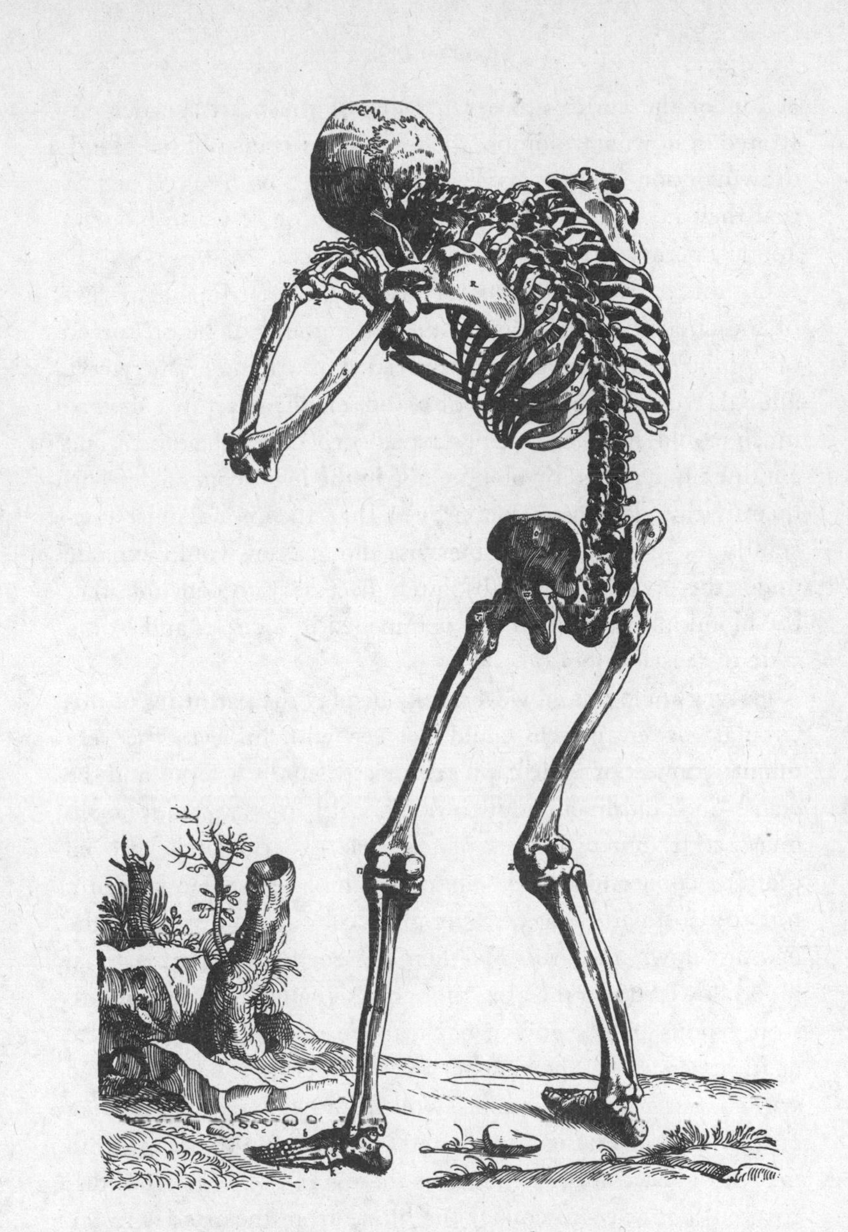

18. Skeleton. Andreas Vesalius, 1543.

action of the auricles and ventricles of the heart, he demonstrated, following Colombo, the pulmonary transit of the blood, drawing upon vivisections he had performed on frogs. (The fact that their hearts beat more slowly than those of warm-blooded animals permitted slow-motion experiments.)

On this basis, in Chapter 8 Harvey announced his discovery of the circulation. He noted that the amount of blood forced out of the heart in an hour far exceeded its volume in the whole animal. Hundreds of gallons of blood left the heart in a day: so much could not conceivably be absorbed by the body and continually replaced by blood made in the liver from chyle. This quantitative demonstration proved that the blood must constantly move in a circuit, otherwise the arteries would explode under the pressure: 'It is absolutely necessary to conclude that the blood in the animal body is impelled in a circle, and is in a state of ceaseless motion.'

Harvey could not, however, fully display the pathways of this circular movement. He could not see with his eyes the very minute connections, the capillaries, between the arteries and the veins – nor did he attempt to do so with the newly developed microscope. But by means of a simple experiment he proved that the connection, albeit unknown, must exist. He ligated a forearm so tightly that no arterial blood could flow below the ligature down the arm. He then loosened it so that arterial blood flowed down the arm, though it remained tight enough to stop venous blood moving back above the ligature. With the ligature very tight, the veins in the arm below it had appeared normal, but now they became swollen, showing that blood must have poured down the arteries and then back up the arm within the veins; hence at the extremities there had to be (as yet undiscovered) pathways to convey the blood from the arteries to the veins.

Finally, Harvey demonstrated that the valves in the veins

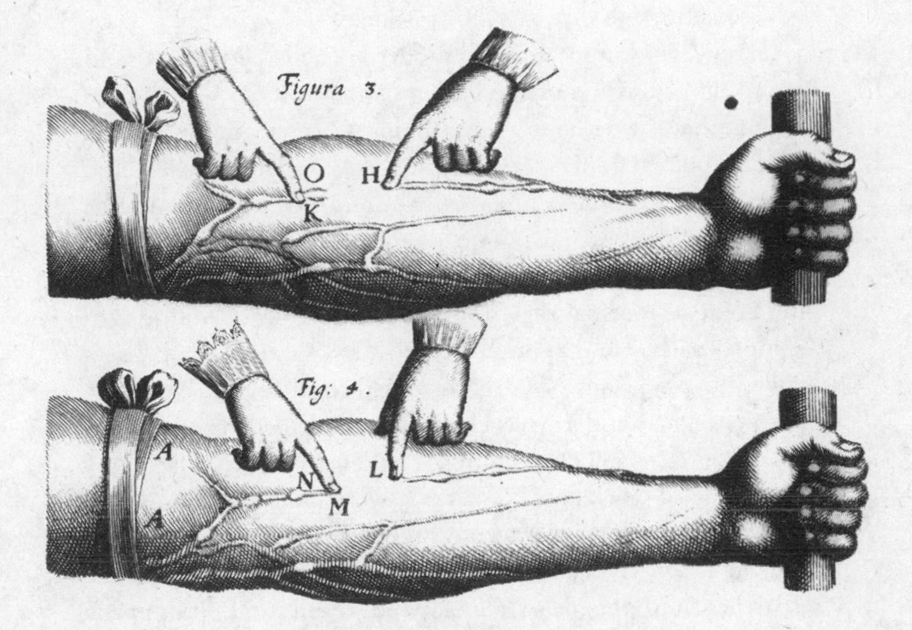

19. Two arms with blood vessels pronounced to indicate circulation. William Harvey, 1628.

always directed blood back to the heart. Contradicting his teacher Fabrizio, he showed that they did not act to prevent the lower parts of the body from flooding with blood. On the strength of the circulation theory, Harvey was able to explain various other previously puzzling phenomena – the rapid spread of poisons through the body, for instance.

Harvey's work appears very modern – he experimented and obeyed the injunction of the Paduan anatomists to see for oneself. But that is true only up to a point. Certainly he looked for himself, but he often saw through Aristotelian spectacles, extolling the perfection of circular motion within the teleological idea of the system of Nature proposed by the Greek father of biology. As so often in Renaissance medicine, innovators built upon, as much as they destroyed, the legacy of Antiquity.

Harvey's new views sparked controversy. Notoriously conservative, the physicians of Paris remained loyal for some time to Galenic dogma, and Harvey himself complained that his own medical practice 'fell off mightily' – patients too were suspicious of newfangled teachings. Nevertheless, his path-breaking work proved a tremendous spur to physiological enquiry.

A clutch of younger English investigators pursued further researches into the heart, lungs and respiration. Prominent among the English Harveians was the anatomist Thomas Willis, remembered for his pioneering study of the brain and of the diseases of the nervous system – to him we owe the term 'neurologie'. The most brilliant, however, was Richard Lower, who studied at Oxford and followed Willis to London. This Cornishman collaborated with the natural philosopher Robert Hooke (of 'Hooke's Law' fame) in experiments that showed how it was the air in the lungs which caused dark red venous blood to be changed into bright red arterial blood. He also earned immortality by conducting the first blood transfusions, transferring blood from dog to dog, and person to person.

These demonstrations took place at the Royal Society, an institution (founded in 1660) which aided the exchange of ideas and techniques between physicians and natural philosophers (later called scientists).

A new investigative aid was the microscope, taken up by Hooke and especially by Antoni van Leeuwenhoek in the Netherlands. Amazing things were discovered: red blood cells, spermatozoa and other micro-organisms. Some believed that tiny little men (homunculi) could be seen in sperm, thus explaining the growth of foetuses.

Influential too were radical conceptual innovations in natural philosophy. As promoted by Descartes, Boyle, Hooke and others, the new or mechanical philosophy proposed the machine as the model for the body. Such mechanists attacked the old scholastic theories, with their talk of virtues and spirits, as pure verbal flimflam, lacking the solid material basis revealed by observation and experiment, and they promoted in their place a hydraulic or hydrostatic understanding of the body's pipes, vessels and tubes, levers, cogs and pulleys. They also prized measurement and quantification. In Padua, Galileo's colleague Sanctorius Sanctorius (1561–1636) developed a thermometer for gauging bodily temperature and the 'pulsilogium', a pulse-watch, and recommended regular weighing of the body to monitor health.

The idea of the 'body machine' boosted research. In Italy, Marcello Malpighi conducted a remarkable series of microscopic studies into the structure of the liver, skin, lungs, spleen, glands and brain. The Pisan Giovanni Borelli and other iatrophysicists (those touting physics as the key to medicine) studied muscle behaviour, gland secretions, heart action, respiration and neural response. Published in 1680, his *De motu animalium* (On Animal Motion) recorded remarkable observations on muscular contraction, the mechanics of breathing, birds in flight, swimming fish, and a host of similar subjects, and interpreted bodily functions primarily in terms of physics. Breathing, for instance, was a purely mechanical process which drove air via the lungs into the bloodstream.

In Borelli's highly innovative work, the physical sciences promised to reveal the secrets of life. His younger contemporary, Giorgio Baglivi, professor of anatomy at Rome, represented the culmination of this iatrophysical programme. 'A human body, as to its natural actions,' he affirmed, 'is truly nothing else but a complex of chymico-mechanical motions, depending upon such principles as are purely mathematical.'

Another innovative attempt to analyse the body in scientific terms lay in iatrochemistry (medical chemistry). Seminal here were the theories of the iconoclastic Swiss doctor, Theophrastus Philippus Aureolus Bombastus von Hohenheim (*c.* 1493–1541), widely dismissed as a quack but respected by some, and also those of his Netherlandish follower, Johannes Baptista van Helmont (1579–1644). 'Paracelsus', as the former liked to style himself – it meant 'surpassing Celsus', the Roman medical authority – replaced the four humours with the three fundamental chemical elements: salt, sulphur and mercury. Questioning Paracelsus's notion of a single *archeus* (or in-dwelling spirit), van Helmont held that each organ had its own individual regulatory *blas* (spirit). His concept of spirit was not mystical but material: all vital processes were chemical, each being due to the action of a ferment or gas capable of converting food into living flesh. Body heat was a by-product of chemical fermentations. Chemistry, broadly understood, was thus the key of life. Such views were radical.

By 1700 advances in gross anatomy and physiology were thus firing hopes of a truly philosophical understanding of the body's structures and functions, cast in the language of the prestigious sciences of mechanics, mathematics and chemistry. The investigations of the following century realized certain of these goals, but also brought frustration in terms of therapeutic payoffs.

In the age of Enlightenment anatomical research continued along Vesalian lines and many splendid anatomical atlases cemented the alliance between art and anatomy. The Dutch anatomist Herman Boerhaave (1668–1738), professor at Leiden and the greatest medical teacher of his day, modelled the bodily system as an integrated, balanced whole in which pressures and liquid flows were equalized and everything found its own level. Spurning the 'clockwork' body of Descartes as too crude, Boerhaave treated it rather as a plumbing network of pipes and vessels, which contained, channelled and controlled body fluids. Health was maintained by the free and vigorous movement of fluids in the vascular system, sickness explained in terms of blockages, constrictions or stagnation. The old humoral emphasis upon balance had thus been preserved but translated into mechanical and hydrostatic idioms.

The presence of some kind of soul was indisputable as a source of animation but, Boerhaave judiciously maintained, to pry into the secret of life was beyond the remit of medicine. The Christian immortal soul was best left to priests and metaphysicians: medicine should study secondary not primary causes, the *how* not the *why* and *wherefore* of the workings of the body.

The founder of the distinguished medical school at the University of Halle, Georg Ernst Stahl (1660–1734), demurred, instead advancing classic anti-mechanistic views. Purposive human action could not be wholly explained in terms of mechanical chain reactions – like balls cannoning round a billiard table. It presupposed, he maintained, the presence of an immaterial soul (*anima*), understood as a presiding and sustaining power in organisms. More than a Cartesian 'ghost in the machine' (that is, one present but essentially separate), the soul for Stahl was the ever-active vehicle of consciousness and physiological regulation, a constant bodyguard against sickness. Friedrich Hoffmann, his younger colleague at the Prussian university, looked more favourably upon the new mechanistic theories of the body. 'Medicine,' he pronounced in his *Fundamenta medicinae* (Fundamentals of Medicine) (1695), 'is the art of properly utilizing physico-mechanical principles, in order to conserve the health of man or to restore it if lost.'

But is the living organism a machine pure and simple, or something more? This was a question put to experimental test in 1712, when the French naturalist René Réaumur demonstrated the capacity of the claws and scales of lobsters to regrow after being severed; in the 1740s, Abraham Trembley divided polyps or hydras, and found that complete new individuals grew. There was obviously more to creatures than the cogs and strings allowed by hard-line Cartesians. The 'nature of life' debate was thus no mere arid armchair speculation: it involved experimental researches into bodies human and animal, putting conjectures to the test. Was digestion performed by some internal vital force? By the chemical action of gastric acids? Or by the mechanical activities of churning and pulverizing by the stomach muscles? Digestion processes were among those subjected to sophisticated experimentation in the eighteenth century.

Experimentation bred new views regarding the nature of vitality – and, by implication, the relations between body and mind (or soul). Here the towering figure was the Swiss polymath Albrecht von Haller, who produced a ground-breaking text, the *Elementa Physiologiae Corporis Humani* (Elements of the Physiology of the Human Body) (1759–66). This included an experimental demonstration that irritability (contractility) is a property inherent in *muscular* fibres, whereas sensibility (feeling) is the exclusive attribute of *nervous* fibres. The *sensibility* of the nerves lies in their responsiveness to painful stimuli; the *irritability* of the muscles is their property of contracting in reaction to stimuli. Haller could thereby advance a physiological explanation – lacking in Harvey – as to why the heart pulsated: it was the most 'irritable' organ in the body, hyperstimulated by the inflowing of blood and responding with systolic contractions. These concepts of irritability and sensibility laid the foundations for modern neurophysiology. Like Newton with gravity or Boerhaave on the soul, he held that the causes of such vital forces were beyond science.

A school of 'animal economy' (the traditional name for physiology) also arose at the impressive new Edinburgh University medical school, founded in 1726. One who built upon Haller's work was William Cullen (1710–90), the university's most eminent professor of medicine. Cullen regarded life itself as a function of nervous power, and emphasized the key role of the nervous system in disease causation, especially mental illness. His follower-turned-foe, John Brown, a larger-than-life figure who died an alcoholic, reduced the whole business of health and disease to variations around Hallerian irritability, though he substituted the idea that fibres were 'excitable'. Animation was to be understood as the product of external stimuli acting upon an organized body. Life, pronounced Brown, was a 'forced condition'; sickness was disturbance of the proper functioning of excitation; and diseases were to be treated as 'sthenic' or 'asthenic', depending on whether the sick body was respectively over- or under-excited. Alcohol and opium in large doses were recommended for either condition – 'Brunonian' medicine had an appealing simplicity.

In France, it was professors at Montpellier University, always more go-ahead than Paris, who headed the vitality debate. Boissier de Sauvages denied that mechanism *à la* Boerhaave could explain the origin and continuation of purposive motion in the body. What was needed was physiological study of the living (*not* dissected) body, endowed as it was with soul. Later Montpellier teachers, notably Théophile de Bordeu, gave vitalism a more materialist spin; addressing the role of physical organization, they discounted an implanted soul and stressed the inherent capacities and energies of organized bodies.

Parallel lines of enquiry were pursued in London. The Scottish-born John Hunter (1728–93), who had trained at his elder brother William's anatomy school, proposed a 'life-principle' to account for the properties which elevated living organisms above gross inanimate matter: this life-force lay in the blood. Thus the bold but simplistic 'body machine' (*machina carnis*) philosophies of the age of Descartes gave way to more dynamic ideas of vital properties or vitalism. It is no accident that the very term 'biology' was coined around 1800.

This new physiology gained much from other sciences emerging out of the scientific revolution. Cullen's contemporary the chemist Joseph Black formulated the idea of latent heat and identified 'fixed air', crucial to the further understanding of respiration as developed by Lower. The French chemist Lavoisier (who, having made a fortune under the *ancien régime* as a tax farmer, lost his head in the Revolution) explained the passage of gases in the lungs. The air inhaled was converted into and exhaled as Black's 'fixed air' – that is, in the Frenchman's new chemical nomenclature, carbon dioxide. Respiration was the analogue within the living body of combustion in the external world: both drew on oxygen and both gave off carbon dioxide and water. It was thus Lavoisier who established that oxygen was indispensable to life.

An enthusiast for this new gas chemistry, the Bristol doctor Thomas Beddoes dreamed of curing many diseases, including tuberculosis, through administrating oxygen and other pure gases to patients. In the process he discovered nitrous oxide (laughing gas) though he failed to follow up its anaesthetic properties (see Chapter 6).

Advances in other sciences also promised medical payoffs, notably experimental electrophysiology, pioneered by Luigi Galvani. In *De Viribus Electricitatis in Motu Musculari* (On Electrical Powers in the Movement of Muscles) (1792), the Italian naturalist described experiments in which the legs of dead frogs were suspended by copper wire from an iron balcony. When this caused the severed limb to twitch, Galvani concluded that electricity was involved – indeed, was integral to the life force. His experiments were extended by Alessandro Volta, professor at Pavia. In his *Letters on Animal Electricity* (1792), Volta demonstrated that muscles could be made to contract by electrical stimuli. The connections between life and electricity revealed by such experiments proved fundamental for neurophysiology. They also provided one of the inspirations for Mary Shelley's 1818 science-fiction fantasy *Frankenstein*, a Gothic account of the creation of life in a man-made monster through physico-chemical means, intended as a cautionary tale of the misuse of Promethean power.

These mechanistic and experimental investigations transformed the thinking about disease. The pursuit of gross anatomy after Vesalius turned attention to the connections between sickness in the living and the pathological signs afforded by the corpse. The growing conviction that post-mortem investigation was the key to the physical changes brought about by disease – not least, the cause of death – was clinched by Giovanni Battista Morgagni (1682–1771), professor of anatomy at Padua. His *De sedibus et causis morborum* (On the Sites and Causes of Disease) (1761) drew on the findings of no fewer than 700 autopsies to demonstrate how bodily organs revealed the footprints of disease.

De sedibus addressed in turn diseases of the head, the chest

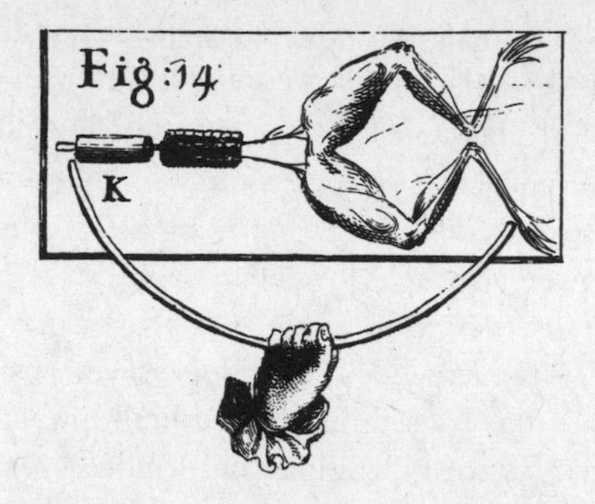
Fig:14
K

Fig:15

Fig:16.

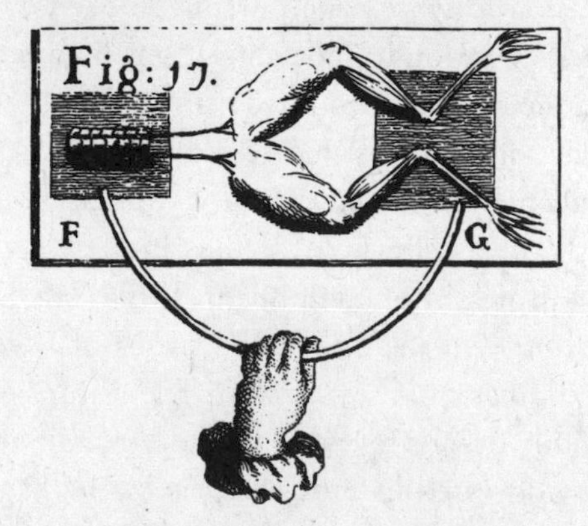
Fig:17.
F
G

and the abdomen. Its case histories, which detail striking symptoms and autopsy results, were followed by an elucidation of the relationships between the case and morbid anatomy. Morgagni's discoveries were numerous. He described the anatomical decay observable in angina pectoris and myocardial degeneration, and the fibrinous clots found in the heart after death; he associated cyanosis (blueness of the skin) with pulmonary stenosis (narrowing of the vessels): and he made major observations on arteriosclerosis (hardening of the arteries).

Morgagni's investigations thus shifted emphasis from the *symptoms* to the *site* of disease – or, to put it another way, he encouraged a shift from a physiological theory (disease is an abnormal condition of the whole organism) to an ontological theory of disease (disease is an entity residing locally in a part). Thinking anatomically, he showed that diseases were located in specific organs, that symptoms tallied with anatomical lesions, and that such morbid organ changes were responsible for disease. Pathology had now been put, alongside anatomy, on a scientific footing.

The outstanding significance of Morgagni's work was recognized and developed by others. Arranged by organs, Matthew Baillie's *Morbid Anatomy of Some of the Most Important Parts of the Human Body* (1793) contains several classic descriptions, including emphysema and cirrhosis of the liver, which he linked to alcohol. By developing in the second edition (1797) the idea of 'rheumatism of the heart' (rheumatic fever), Baillie contributed to the early study of heart disease.

Pathology's next milestone was the publication in 1799 of the *Traité des Membranes* (Treatise on Membranes) by Marie François Xavier Bichat. A doctor's son from the Jura, Bichat

20. (opposite) *Experiments on animal electricity using frogs' legs. Galvani, 1791.*

settled in Paris and became assistant to the leading surgeon, Pierre-Joseph Desault. Concentrating on structures comparable in texture but present in different organs, he described twenty-one such tissues, distinguished by appearance and vital qualities. The most widespread were cellular tissue, nerves, arteries, veins, absorbent and exhalant vessels. Just like the elements in Lavoisier's new chemistry, these tissues were, for Bichat, the analytical building-blocks of anatomy, physiology and pathology, and he set about delineating their structure, vital properties, responsiveness and abnormalities. Henceforth, he claimed, diseases must be seen as lesions of specific *tissues* rather than (as for Morgagni) of *organs*. 'The more one will observe diseases and open cadavers,' he declared, 'the more one will be convinced of the necessity of considering local diseases not from the aspect of the complex organs but from that of the individual tissues.'

By thus viewing pathology with fresh eyes Bichat laid the foundations for nineteenth-century clinical medicine. And, as will be seen in the next chapter, the pathological anatomy for which Paris became renowned did not merely build upon his tissue pathology, it heeded his directive. 'You may take notes for twenty years from morning to night at the bedside of the sick,' he taught, 'and all will be to you only a confusion of symptoms . . . a train of incoherent phenomena.' But start cutting bodies open and, at a stroke, 'this obscurity will soon disappear'. Here was the medicine of the all-powerful gaze, one which saw – almost with X-ray eyes – through the patient to the underlying disease. The anatomizing eye was pressing on still further.

CHAPTER FOUR

The Laboratory

Chance favours the prepared mind.
Louis Pasteur

The Renaissance, as we have seen, launched the new science and the Enlightenment blew the trumpet for it, but it was the nineteenth century which was the first true age of public science, funded by the state and promoted by universities and research institutes. For the first time, it made good sense for an ambitious young medical man, like Dr Lydgate in *Middlemarch*, to acquire a scientific grounding and to wear a scientific air – though, as George Eliot's hero found, the community's response could be decidedly double-edged.

Lydgate had significantly been trained in Paris. Around 1800 medical investigation and thinking were revolutionized by a clutch of Paris physicians, who availed themselves of the opportunities afforded by the centralization brought about by the Revolution to use big public hospitals to advance medical science. The best remembered is René Théophile Hyacinthe Laënnec (1781–1826), pupil of Bichat, physician to the Salpêtrière Hospital and the Hôpital Necker, and inventor of the stethoscope (1816).

Initially a simple wooden cylinder, about nine inches long and with a single earpiece, the stethoscope proved the key diagnostic innovation, at least until the discovery of X-rays in 1895. Gaining expertise thereby in breath sounds both normal and abnormal, Laënnec diagnosed, and gave outstanding clinical and

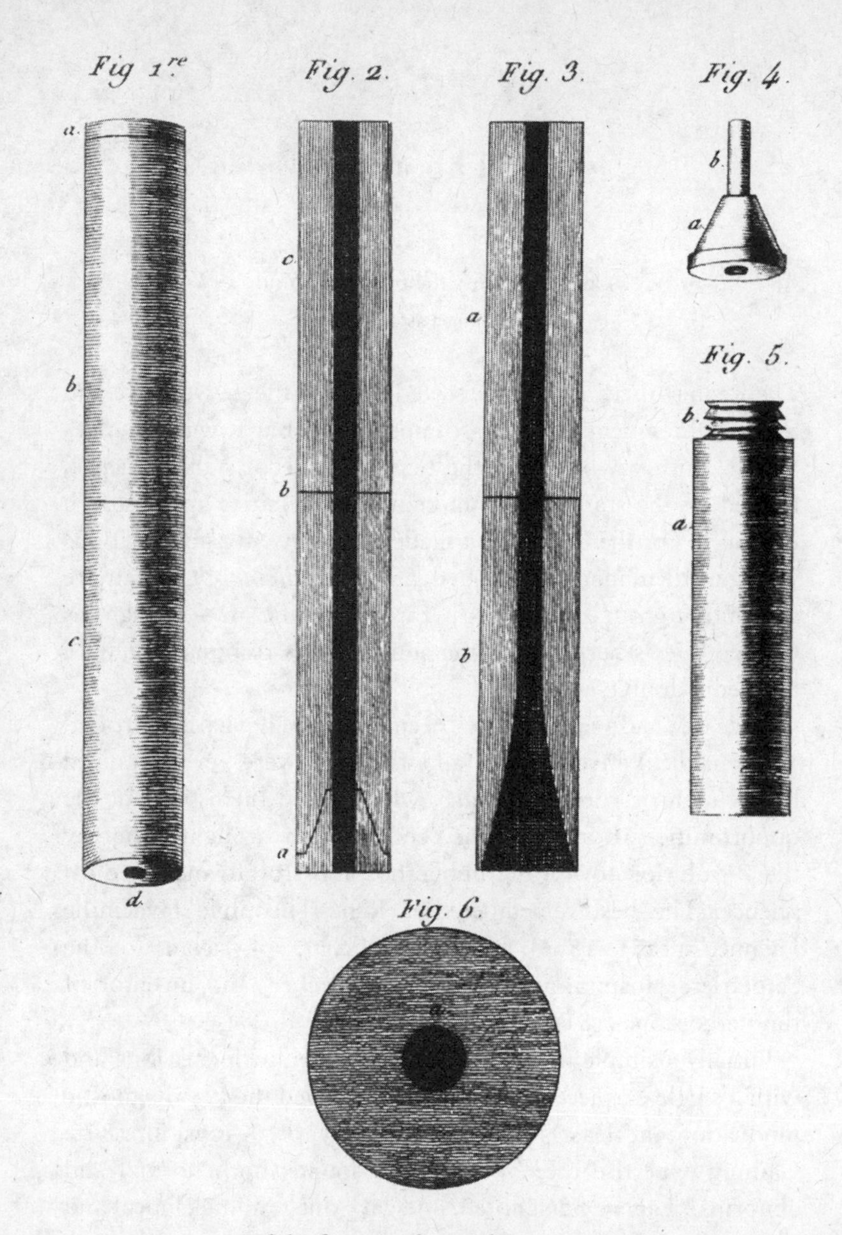

21. *Diagram of the first wooden stethoscope. Laennec, 1819.*

pathological descriptions of, a range of pulmonary ailments: bronchitis, pneumonia and above all tuberculosis (phthisis or consumption). Stethoscopy became standard practice over the next decades as translations of his writings publicized the technique and a stethoscope draped around the neck became the enduring icon of modern medicine: it had the word science blazoned all over it.

Hardly less influential was Pierre Louis, who, alongside a massive book on tuberculosis (1825) and another on fever (1829), spelt out the agenda of the new hospital medicine in his *Essay on Clinical Instruction* (1834). This held that *symptoms* (that is, what the patient felt) were secondary in clinical value; far more significant were *signs* (what physical examination found). On the basis of such signs, the lesions of the diseased organs could be determined – these were the most objective guides to identifying disease, making prognoses and, where feasible, devising remedies. (Endlessly exposed to the dying poor, the Paris doctors rated diagnostics above therapeutics.)

Clinical medicine was thus, for Louis and his colleagues, an observational science to be learned on the hospital ward and in the morgue through the recording and explication of facts. Medical training must be a discipline of the explanation of the sights, sounds and smells of disease – an education of the senses. Clinical judgement, the doctor's true metier, lay in astute interpretation of what experienced senses perceived.

Louis was, furthermore, a passionate advocate of arithmetical methods designed to test therapies numerically – initiating what would later be called clinical trials. The scale of the Paris hospitals allowed these doctors to transcend the individual case for statistical probabilities.

Laënnec, Louis and their peers and followers meticulously delineated pathological signs in the living and the dead alike. The shift this brought from symptoms (variable and subjective) to signs (constant and objective) helped ensconce the concept that diseases were discrete entities – *real things* – and taught that diseased states were different in kind from normal ones. This is known as the ontological theory of disease.

Not all agreed. Espousing strikingly different views on the relations between the normal and the pathological, another Paris physician, F. J. V. Broussais, accused his colleagues of perverting medicine with their localizing pathological anatomy, their dogmatism about disease specificity, and their therapeutic gloom. The true legacy of the brilliant Bichat (see Chapter 3) lay, argued Broussais, in his grasp of the primacy of physiology, and hence of the continuum between the normal and the pathological. Sickness was not fundamentally and qualitatively different from health; rather sickness set in when normal functions went awry. Such thinking was later critical for Claude Bernard and Rudolf Virchow (see below).

Students from North America and Europe flocked to Paris and returned home beating the drum for French medicine, armed with skills in pathology, chemistry and microscopy – and a stethoscope in their valise. Medical education everywhere grew hospital-based and more systematic. Inspired by Paris-trained teachers, London medicine experienced a boom: by 1841 St Bartholomew's Hospital had 300 pupils, and from the 1830s the capital had also boasted a teaching university, with two colleges, University and King's, each with medical faculties and purpose-built hospitals.

Vienna shone in particular. There the Paris-inspired Carl von Rokitanski (1804–78) made pathological anatomy compulsory – indeed, almost a fetish. The age's most obsessive dissector, supposedly performing 60,000 autopsies in all, Rokitanski displayed an unparalleled mastery of anatomy and pathological science, and left notable studies of congenital malformations, pneumonia, peptic ulcer and valvular heart disease. Till the First

World War, Vienna remained one of Europe's supreme medical (to say nothing of psychiatric) centres.

Thanks to the Paris school, the hospital became a nodal point for medical investigation: its wealth of clinical material was unsurpassed. A rival research institution developed alongside: the laboratory. By 1850 laboratories were transforming physiology and pathology and making their mark too upon medical education.

Laboratories were far from new – they were an innovation of the age of Boyle and Hooke; nor, for that matter, was experimental medicine. Nevertheless nineteenth-century practitioners of organic chemistry, microscopy, physiology and other medicine-related disciplines were right to believe they were in at the birth of a new enterprise: while the hospital, they conceded, was fine for making observations, the laboratory was the place for systematic controlled experimentation.

German universities in particular promoted the research ethos, and Justus von Liebig's Institute of Chemistry at the University of Giessen set the mould for laboratory science. Liebig (1803–73) developed an influential programme for subjecting living organisms to strictly quantified chemical analysis. By measuring and analysing what went in (food, oxygen and water) and what came out (urea, salts, acids and carbon dioxide), vital evidence would be revealed about what would later be called internal metabolic processes.

The body, held Liebig, was an ensemble of chemical systems. Respiration brought oxygen into the body, where it mixed with starches to liberate energy, carbon dioxide and water. Nitrogenous matter was absorbed into muscle tissues; urine was the ultimate waste product, together with phosphates and assorted other chemical by-products. Chemical analyses of blood, sweat, tears and urine were undertaken, so as to quantify the equations in living organisms between food and oxygen consumption and energy production. Launching systematic investigation of nutrition and metabolism, Liebig and his school thus inaugurated what was to be called biochemistry.

Liebig trained up students by the squadron in organized laboratory research, dedicated to applying the models and methods of the physical sciences to living organisms. As early as 1828 his friend Friedrich Wöhler, from 1836 professor of chemistry at Göttingen, synthesized the organic substance urea from inorganic substrates – convincing proof that no categorical barrier separated the vital compounds found in living beings from ordinary chemicals. Such findings gave impetus to the reductionist ethos which ridiculed the speculative, idealistic philosophy of the Romantics with their mystical aspirations to fathom the meaning of life. Scientific materialism became the dominant philosophy in the German research schools of the second half of the century.

Physiology came of age as a high-status experimental discipline. Its trailblazer was Johannes Müller, from 1833 professor of physiology and anatomy at Berlin. His enormous *Handbook of Physiology* (two volumes, 1833–40) served for many years as the bible of the discipline. He was an inspiring teacher and his students – Theodor Schwann, Hermann von Helmholtz, Emil du Bois-Reymond, Karl Ludwig, Ernst Brücke, Jacob Henle, Rudolf Virchow and many others – became the directors of scientific and medical research in the German world and enjoyed international reputations.

Four of Müller's protégés – Helmholtz, du Bois-Reymond, Ludwig and Brücke – published a manifesto in 1847 proclaiming that physiology's goal was to explain all vital phenomena reductionistically, in terms of physico-chemical laws. With this commitment to scientific naturalism, experimental physiology aimed, as Ludwig put it, to understand functions 'from the elementary conditions inherent in the organism itself': what was the stuff of life and how was it organized?

Helmholtz devoted himself to the measurement of animal heat and the velocity of nerve conduction, and developed the ophthalmoscope, to aid work on vision. Ludwig for his part conducted pioneering research on glandular secretions, notably the manufacture of urine by the kidneys. Du Bois-Reymond, professor of physiology in Berlin, mainly pursued electrophysiological studies of muscles and nerves. Brücke went to Vienna, where his concerns spanned physiological chemistry, histology and neuromuscular physiology. He was one of Freud's teachers and heroes.

As well as the sacrifice of animals in vivisection experiments, such research required improved measuring and data-recording instruments. In 1847 Ludwig introduced the key device, the kymograph, a multi-purpose apparatus designed to trace bodily alterations – for instance, the pulse – on to a line on a graph. Technological sophistication became integral – indeed, indispensable – to medical science.

The microscope was also greatly improved from around 1830 through the correction of distortion, enabling rapid progress to be made in the new science of histology, the microscopic study of tissues. Advanced microscopy made possible the revolutionary new science of cells (cytology), initiated in 1838 by another of Müller's pupils, Theodor Schwann, who extended cell theory, restricted since Hooke to plants, to animal tissues. Schwann proposed a reductionist model of these ultimate capsules of life. Cells were the fundamental units of zoological and botanical activity; they incorporated a nucleus and an outer membrane; and, somewhat like crystals, they were formed out of an amorphous matrix, the 'blastema'.

Schwann's views on the materiality of the blastema were challenged by Rudolf Virchow, professor of pathological anatomy at Würzburg in 1849 and Berlin in 1856, and surely the most creative German medical researcher of his times. He advanced the maxim: *omnis cellula e cellula* (all cells come from cells). If Bichat put tissues on the map, Virchow did the same for cells. In his hands cell theory was credited with huge explanatory power for biological events such as fertilization and growth, and for pathological ones, as, for instance, the source of pus in inflammation. Cancer arose, he brilliantly showed, from abnormal changes within cells which then multiplied out of control through division (metastasis). In the study of cells lay the key to the understanding of disease. Virchow thus espoused an *internal* concept of disease; partly for that reason he later proved leery of Pasteurian bacteriology, which he regarded as rather superficial because it saw disease as essentially *external* in its causation. (Fierce Franco/German rivalry also played its part.)

From the 1850s German laboratories attracted students from all over Europe and North America while France, recently the leader, slipped behind, as it failed to create the laboratories necessary for state-of-the-art physiological research. Nevertheless France continued to produce eminent researchers, above all Claude Bernard (1813–78).

After failing in his dream of becoming a dramatist, the young Bernard opted for medicine. Success followed success, including a chair at the Sorbonne, a seat in the Senate and the presidency of the French Academy. He was responsible for major physiological demonstrations: the effect of such poisons as carbon monoxide and curare on the muscles; the role of the liver in maintaining blood glucose levels; the digestive functions of the secretions of the pancreas; and the role of the vasodilator nerves in regulating blood-flow in blood vessels, to name just a few. Above all, he set out an agenda for the biomedical sciences in his 1865 classic, *Introduction to the Study of Experimental Medicine*.

Hospital medicine *a là* Laënnec, argued Bernard, had serious limitations: like natural history, it was passive, and the sickbed presented too many imponderables. To achieve progress in physiology required the active involvement of the experimentalist, under strictly controlled conditions. Moreover – and here he leant more towards Broussais than to the 'Paris school' – the pathological lesion itself was not the origin but the accompaniment or the consequence of disease. Pathophysiological knowledge could be gained only in the laboratory through vivisection experiments performed on laboratory animals in managed environments. The interplay of physiology, pathology and pharmacology constituted the key to experimental medicine, and each had to be a laboratory science.

Yet Bernard was no vulgar materialist or reductionist. Living creatures were not automata wholly at the mercy of the external environment, for higher organisms did not live solely in that milieu – they created their own internal environment. Physiological mechanisms were dedicated to balancing the sugar, salt and oxygen concentrations of the blood and tissue fluids; it was their job to preserve uniform body temperature in the face of external fluctuations. It was through these equilibrating mechanisms – later called 'homeostasis' – that higher organisms achieved some autonomy within the law-governed determinism of the natural order. Bernard's insights informed all later researches into the normal and the pathological.

Scientific medicine emerged more slowly in Britain and the USA, though increasing numbers of students from these nations went to German universities to study biology and medicine. One of them, William Henry Welch, injected German methods into American experimental medicine at the most Teutonic of American campuses, Johns Hopkins in Baltimore, where he became professor in 1878. Its medical school – singular in those times for admitting women – prized advanced teaching and research. Also highly significant was the opening in 1901 of the Rockefeller Institute for Medical Research in New York – it proved the nursery of many a later Nobel Prize winner.

In mid-Victorian Britain, medicine remained chiefly in the hands of private practitioners, and the universities got meagre state support for research. Nor were the prospects for medical research helped by noisy public hostility to experimentation. Anti-vivisection campaigns led to the 1876 Cruelty to Animals Act, a compromise which permitted medically qualified investigators to conduct vivisection experiments only under licence and in strictly stipulated conditions. No other nation passed matching legislation before the twentieth century.

Gradually, however, British physiology won its place in the sun. Working first in London and then in Edinburgh, Edward Shäfer (later Sharpey-Shäfer) won fame for his researches on muscular contraction, while Michael Foster and his pupils J. N. Langley and W. H. Gaskell created a research school in Cambridge which produced a crop of future Nobel Laureates, including Henry Dale and Lord Adrian.

The superstar of the next generation of medical researchers, Louis Pasteur (1822–95), was, oddly, no physician, but a chemistry graduate of the Ecole Normale Supérieure in Paris. He was an outstanding microscopist, whose interest in micro-organisms was stirred by studies of fermentation in connection with wine- and beer-making, and he devised elegant experiments to scotch the old theory of spontaneous generation. Maggots, he showed, arose from insect-laid eggs and from organisms pervading the atmosphere, invisible to the naked eye. On this basis he developed his acclaimed method for eliminating microbes from milk: 'pasteurization' – heating to a prescribed temperature to kill them – ensured that milk would cease to be a source of tuberculosis and gastro-enteric ailments.

22. *Pasteur in his laboratory among various pieces of scientific equipment.*

Controversy as to what causes disease – the problem of aetiology – was one of medicine's key unresolved questions, and it was brought to a head by the terrible waves of epidemics blighting industrializing Europe. Many espoused the 'miasmatic theory' – the idea that disease originated in effluvia and other emanations from the soil and atmosphere. Others embraced 'contagionism' – disease was something passed from person to person. There was a multitude of variants on and combinations of such views, and none held sway.

Pasteur by no means invented the 'germ theory' – disease is caused by invasion of the body by microscopic living organisms: it had long been touted. But he was the first to show, through convincing experimental demonstrations, that particular microbes actually caused particular diseases – in cattle, pigs, poultry and, finally, humans.

And, being of a practical rather than a theoretical bent, he turned himself to the therapeutic potential of the germ theory. His researches into chicken cholera, swine erysipelas and anthrax led to new 'vaccines' – the term he coined to honour Edward Jenner, the English country doctor who, at the close of the eighteenth century, had championed cowpox inoculation against smallpox (*vacca* is Latin for cow).

The efficacy of Pasteur's anthrax vaccine was shown in one of the many spectacular experiments which were his forte. On 28 April 1881 he injected twenty-four sheep with his new vaccine, repeating it after three weeks. A further fortnight later this group, along with a control group of unvaccinated animals, was implanted with virulent anthrax bacilli. When the sheep were again inspected on 2 June, all the vaccinated animals were healthy, whereas all the unvaccinated ones were dead or dying. Pasteur's crowning achievement, the rabies vaccine he developed in 1885, was for a ghastly and fatal disease which, like anthrax, affected both animals and human beings.

23. Edward Jenner in the Smallpox and Inoculation Hospital. Etching, James Gillray, 1801.

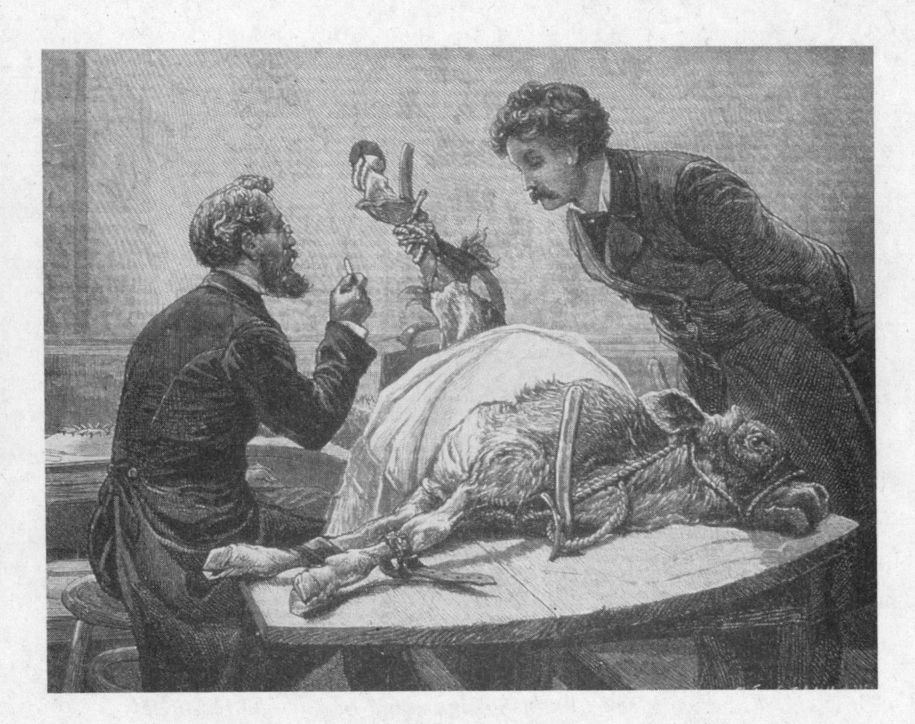

24. *Taking lymph from the calf to create vaccinations.*
C. Staniland, 1883.

Pasteur's linkings of various streptococci and staphylococci to specific diseases put bacteriology on the scientific map. It was, nevertheless, his younger German contemporary, Robert Koch, later professor of public health in Berlin, whose meticulous demonstrations ultimately clinched the microbial theory of disease causation and gave it theoretical solidity.

Trained by Wöhler, in 1879 Koch published a paper, 'The Aetiology of Traumatic Infectious Diseases', which proved a milestone in the methods of medical science. It differentiated between distinct bacteria, connected specific micro-organisms to specific infections, and sought to prove that bacteria were the cause of infections. To this end he spelt out what have ever since been known as 'Koch's Postulates'. To prove a particular micro-organism produces a particular condition, four requirements must be satisfied:

(1) The specific organism must be present in every instance of the infectious disease;
(2) the organism must be capable of cultivation in pure culture;
(3) inoculating an experimental animal with the culture would reproduce the disease; and
(4) the organisms could be recovered from the inoculated animal and grown again in a pure culture.

As lately with AIDS, 'Koch's Postulates' are still invoked in attempts to test whether a specific micro-organism is the true (necessary and sufficient) cause of a disease.

Koch's greatest concrete discoveries were the bacillae which produce tuberculosis (1882) and cholera (1883). His students – and rivals – went on to use his methods to identify the causal microbes for typhoid, diphtheria, pneumonia, gonorrhoea, undulant fever, meningitis, leprosy, tetanus, plague, syphilis, whooping cough, and many other staphylococcal and streptococcal infections. By thus highlighting living pathogens, the microbe hunters who headed the new bacteriology made great strides towards solving the thorny problem of disease aetiology, though in the process throwing up the perplexing questions of susceptibility and resistance, which proved the matrix for the later science of immunology.

Microbiology was crucial to developments in tropical medicine. A product of the needs and opportunities of political, military and economic imperialism, that specialty played a key part in the global spread of Western power. Medicine followed trade and the flag. A direct response to colonialism, it did not merely expedite such expansion but provided a justification for it: was it not part of the white man's mission to bring medicine to the deadly tropics? All too often it was conveniently overlooked that the white man was in large measure responsible for making them so unhealthy in the first place.

Fatal experience taught that the tropics were the white man's grave, and trade and imperial designs had long been checked by diseases like yellow fever and malaria (*mal aria*, bad air). Yet the relations between climate and disease were hotly contested. Traditional explanations for the maladies of the tropics had drawn upon a miasmatic environmentalism ultimately rooted in Hippocratic teachings: heat produced putrescence (rotting vegetation, etc.) which gave off bad airs (miasmata) that bred terrible fevers. Challenging alternative explanations emerged in the last quarter of the century; their pioneer was the Scot Patrick Manson.

Manson served from 1866 as a Customs Medical Officer in Amoy (Hsai-men), off the south-east China coast. Studying elephantiasis, a chronic disfiguring disease marked by massive swelling of the genitals and limbs, he showed that it was caused by a parasite – *Filaria*, a nematode worm – spread via mosquito bites. This was the first disease proved to be transmitted by an insect vector. It became a powerful explanatory model.

Building a reputation as a parasitologist, Manson stamped his vision upon the emergent specialism. Assimilating the new bacteriology, he indicted parasitic organisms as the agents of tropic diseases. Over the next generation schistosomiasis (see Chapter 1) was found to be produced by a worm, the trematode *Bilharzia*; tropical dysentery by an amoeba; sleeping sickness, that appalling African affliction, by a *Trypanosome*, a protozoan; and malaria by another sort of protozoan, the *Plasmodium*.

The great scourge of malaria was cracked by Ronald Ross (1857–1932), a member of the Indian Medical Service. In 1894 Manson hinted to his junior colleague that it was transmitted through mosquito bites; Ross went to work on his hypothesis. Duplicating earlier investigations by the French microbiologist Charles Laveran, he discovered the malaria parasite in the stomachs of Anopheles mosquitoes which had bitten malaria sufferers. He then showed that the mosquito was a necessary vector in transmission, by elucidating the relationship between the *Plasmodium* life-cycle and disease (see Chapter 1). The Italian Giovanni Grassi independently traced malaria to mosquitoes, but it was to Ross alone that the Nobel Prize was awarded in 1901.

Other diseases also succumbed to this parasitological model. The shocking mortality from yellow fever in the Spanish–American War in Cuba (1898–1901) led to the setting up in 1900 of a US Army Yellow Fever Commission, headed by Walter Reed from Johns Hopkins University and James Carroll of the US Army Medical Corps. A Havana doctor, Carlos Finlay, had earlier outlined a mosquito-borne theory of yellow fever based on experiments in which healthy volunteers were bitten by mosquitoes which had fed on yellow fever victims;

they then typically fell sick. Finlay was proved correct by the Americans, though this time a different species of mosquito was responsible, *Aedes aegypti*. A successful mosquito eradication programme was conducted in Havana, and a similar strategy was followed in the Panama Canal Zone, where the French plan to build a canal had been scuppered by appalling yellow fever losses. Draining marshes and reducing stagnant water led to a major decline in mosquito-borne diseases, and construction of the Panama canal finally went ahead between 1904 and 1914 – an outstanding public triumph for medical science.

Such victories left an ambiguous legacy, however. They encouraged the arrogant belief that the health problems of the tropics could readily be solved by a dose of Western medical scientific intervention. The failure of twentieth-century campaigns to eradicate malaria (among others) by such methods attests the fallacy of such thinking. All too often approaches and investments in medicine proved quite inappropriate to true Third World needs.

The twentieth century was to bring countless breakthroughs in biology, chemistry, physiology, as well as a proliferation of many new specialties, all under the umbrella of medical science. There was no one single royal road to success. Some developments drew on lucky accidents (as with penicillin); others were the fruits of indefatigable investigation: Paul Ehrlich tested over 600 arsenicals before he hit upon Salvarsan (for both, see Chapter 5) against syphilis. Different disciplines linked up with each other in quite unforeseeable ways, throwing unexpected light upon the workings of the body and of disease. In range and number the triumphs of twentieth-century medical science were unparalleled, bringing myriad spectacular transformations in practical medicine. These are far too numerous to chronicle here, but a few of the most signal developments will be mentioned.

The microbiological revolution launched by Pasteur and Koch brought into being the key new science of immunology. What explains the resistance, natural or artificial, of a host? How should medicine capitalize upon it? The Frenchman had pointed to nutritional factors in immunity, but he was more concerned with vaccine development than with immunological theory. In 1884, however, the Russian Elie Metchnikoff observed a phenomenon he termed 'phagocytosis' (cell-eating): amoeba-like cells in lower organisms apparently had the power to ingest foreign substances. Might not these cells be akin to the pus cells visible in higher creatures? Microscopic examination of animals infected with anthrax and other pathogens showed white blood cells targeting and seemingly digesting the disease germs – it was like an army that was fighting infection.

Metchnikoff's cellular theory of antigens, antibodies and resistance won over the French scientific community, but (almost predictably) German researchers came up with a rival idea: chemical therapies. Emil von Behring and Paul Ehrlich argued that immunological warfare was waged less by the white blood cells than by the blood serum. Working with the Japanese researcher Shibasaburo Kitasato, von Behring announced in 1890 that the blood serum of an animal, rendered immune to tetanus or diphtheria by the injection of the relevant toxin, could treat another exposed to an otherwise fatal dose of the bacilli. This became known as 'serum therapy' and it enjoyed some success, and antitoxins went into production for tetanus and diphtheria, pneumonia, plague and cholera.

Many aspects of the immunological response remained puzzling, but the work of the Australian MacFarlane Burnet and other immunologists in the 1950s and 1960s at last elucidated the mechanisms of the production of antibodies in the human body in a synthesis which showed the unity of the nervous and endocrine systems with the immune system. The science leapt to public prominence from the 1980s thanks to AIDS, a disease which brought about the destruction of the natural immune system.

Questions of susceptibility and resistance obviously bore upon understanding of the relations between nutrition and sickness. The ancient problem of scurvy on oceanic crossings had bred conjectures connecting disease to diet. As early as 1747 James Lind, physician at the Royal Naval Hospital at Haslar, conducted the first classic therapeutic trial. He divided a dozen scurvy sufferers into six groups of two, and treated each pair with a different remedy. Those given two oranges and a lemon each day recovered best. Lind's work induced the Admiralty to supply the navy with lemon juice, the result being that during the Napoleonic Wars the British navy suffered far less from scurvy than the French.

The researches of Liebig, discussed above, set the organic chemistry of nourishment and digestion upon a sound footing. Exploring the creation of energy out of food, his studies established the ideal of a balanced diet. Transcending the proven links between sickness and starvation, however, a new concept was emerging around 1900: deficiency disease, the idea that a healthy diet required very specific chemical components. Crucial were Christiaan Eijkman's investigations into beriberi (with its classic symptoms of muscular weakness and dropsy), which led him to propose the concept of 'essential food factors', or roughly what would in 1912 be dubbed 'vitamins' by the chemist Casimir Funk. Through clinical studies on prisoners on Java, then a Dutch colony, Eijkman showed that the substance (now known as vitamin A) which gives protection against beriberi is contained in the husks of rice grains – precisely the element removed when it is polished for the kitchen.

Eijkman's ideas were taken forward by the Cambridge

biochemist Frederick Gowland Hopkins (1861–1947), who similarly discovered that very small amounts of accessory food factors were needed for the body to utilize protein. Research followed into the special functions of different vitamins. In 1928 Albert von Szent Györgi isolated vitamin C, which became recognized as the element in lemon juice effectual against scurvy. The model of deficiency disease proved highly fruitful. In 1914 Dr Joseph Goldberger of the US Public Health Service showed that pellagra, with its classic pot-bellied symptoms, was not, as believed, an infectious disorder but was due to poor nutrition.

Study of nutrition was an extension of the research programme into the internal environment launched by Claude Bernard. So too was endocrinology, or investigation of internal secretions. One of the fruits of the energetic research programme into proteins and enzymes pursued at University College London by William Bayliss and Ernest Starling around 1900 was the key concept of the hormone (from the Greek for 'I excite'). It pointed to a new field: study of the regulatory chemical messengers travelling from particular organs (ductless or endocrine glands) to other parts of the body via the bloodstream.

The thyroid, pancreas, sex glands and the adrenals all became recognized as endocrine glands, essential regulators of health. Once it was discovered that the islets of Langerhans in the pancreas released a material controlling blood sugar level, recognition dawned that diabetes, then fatal, was a hormone deficiency disease. In the race to extract this active substance, victory went to two Canadian researchers, Frederick Banting and Charles Best. On 11 January 1922 they gave the first insulin injections to a fourteen-year-old boy who was dying of diabetes: almost immediately his blood sugar level fell. A critical disease could thereafter reliably be controlled (though not cured).

Further endocrinological researches led to the isolation of the female sex hormone, oestrone. By the 1930s the family of the oestrogens had been elucidated, as had the male sex hormone, testosterone. Twenty years later, on the basis of these discoveries, Gregory Pincus and Carl Djerassi developed an oral contraceptive for women. Launched in 1959, the Pill was the first ever fully effective contraceptive, and it pointed to a new era of lifestyle drugs: ones designed not to counter sickness but to improve living itself. Viagra (1998) for treating male impotence is another.

Experimental neurophysiology also made great strides during the nineteenth century, leading to the English scientist Charles Sherrington's demonstration that the operation of the brain cells involved two neurons with a barrier (the synapse) between which impulses could jump. But how were the nerve currents transmitted from nerve to nerve, across synapses, to their targets? Evidence mounted that chemical as well as electrical processes were at work. In 1914 the English physiologist Henry Dale found a chemical in ergot – he called it 'acetylcholine' – which proved responsible for the transmission of nerve impulses across some kinds of synapses – the first neurotransmitter to be identified.

Seven years later the German physiologist Otto Loewi showed that the heart, when stimulated, emitted the enzyme cholinesterase, a chemical inhibitor which interrupted the acetylcholine stimulator. Numerous other chemical agents were found active in the nervous system, including adrenaline, identified by the Harvard physiologist Walter Cannon and, later, noradrenaline, dopamine and serotonin. The transmitter–inhibitor pattern thus revealed opened up the possibility of controlling or correcting neurophysiological and even psychiatric disorders, while the paralytic action of tetanus and botulism on the nervous system could at last be explained.

Similarly Parkinson's disease, a degenerative nervous condition, was both mysterious and untreatable until it was linked to chemical transmission in the nervous system. In the late 1960s it was found that it could be alleviated with L-dopa, a drug which enhances dopamine in the central nervous system. Research in neurotransmission has thus led to various treatments. Introduced in 1987, Prozac, a drug which by raising serotonin levels creates a feelgood sense of security and assertiveness, started to be prescribed for depression; within five years, eight million sufferers had taken that designer anti-depressant, said to make people feel 'better than well'. Once again, basic biological research in time bore practical medical fruit.

Perhaps outstanding in its premiss among the many other fields of biomedical research has been genetics. The theory of evolution by natural selection advanced in 1859 in the *Origin of Species* inevitably highlighted the role of inheritance in human development and disease. But Darwin hesitated between alternative theories of inheritance, and the false trails of degenerationism and eugenics (see Chapter 8) had great and sometimes lethal consequences – as in the Nazi death camps – before genetics was put on a sound footing.

The real breakthrough for medicine came when the new field of molecular biology led in 1953 to the announcement of the double helical structure of DNA (deoxyribonucleic acid: the building-block of all living material) by Francis Crick and James Watson. What came to be known as the cracking of the genetic code in turn led to the ongoing Human Genome Project, with its goal of mapping all human genetic material in the ultimate Book of Man. Meanwhile, the interplay of clinical studies and laboratory research was establishing the genetic component in such disorders as cystic fibrosis and Huntington's chorea (shown to run in families as long ago as 1872 by the American physician George Huntington but not explained).

Genetics has held out a promise to the sick in three ways. Genetic engineering (biotechnology) has been developed as a means for producing new types of drugs – for instance, human insulin. Genetic screening can be used for diseases such as Huntington's chorea and cystic fibrosis. And, curatively, gene therapy may prove a way to eradicate faulty genes. Through inserting a normal copy of an abnormal gene into a cell, doctors hope that, by generating the correct rather than the scrambled genetic message, genetic disorders might be cured. Opinion, nevertheless, remains divided as to how far such widespread, baffling and terrible diseases as cancers and schizophrenia may be illuminated by geneticists and rectified by genetic engineering. As highlighted by the fraught issues of gene patenting and human cloning, genetic engineering is one issue about which the public feels justifiably anxious regarding the risk of the Frankenstein-like abuse – or at least precipitate deployment – of new biomedical powers.

It was the boast of Hippocratic medicine that it would 'do no harm'. Thanks to the experimentalism made possible by the laboratory, modern medicine grew Promethean in its vision: there was no forbidden knowledge, all things were possible in its mechanical medical model. But the power to do good is double-edged. The fear today is that a 'can do, will do' mentality will prevail at the frontiers of research, clinical medicine and surgery, regardless of wider ethical responsibilities. And the biomedical model can be myopic, searching ever-more microscopically for disease but often omitting the wider picture of populations, environments and health.

CHAPTER FIVE

Therapies

> Throw out opium . . .; throw out a few specifics . . .; throw out wine, which is a food, and the vapors which produce the miracle of anaesthesia, and I firmly believe that if the whole materia medica, *as now used*, could be sunk to the bottom of the sea, it would be all the better for mankind, – and all the worse for the fishes.
>
> Oliver Wendell Holmes, *Medical Essays* (1891)

Organized laboratory investigation in the nineteenth century provided the seedbed for new biomedical sciences. It also became the crucible for dazzling pharmaceutical breakthroughs. This was especially welcome because, as we saw in Chapter 2, therapeutics had lagged behind other branches of medicine – hence the fatalistic doctrine of 'therapeutic nihilism'.

The sheet-anchor of medicine both domestic and professional has always been an array of herbal remedies: leaves, roots, bark, ground up, steeped, made into infusions, etc. The Egyptian Ebers papyrus, for example, recommends: 'To drive away inflammation of the eyes, grind the stems of the juniper of Byblos, steep them in water, apply to the eyes of the sick person and he will be quickly cured.' The Greeks Theophrastus (fourth century BC) and Dioscorides (first century AD) compiled herbals and accounts of *materia medica*, dealing with aromatics such as saffron, oils, salves, shrubs and trees. Arab medicine added new preparations. The medical formulary of al-Kindi (*c.* 800–870), for instance, contained many Persian, Indian and Oriental drugs quite unknown to the Greeks, including camphor, cassia, senna, nutmeg and mace, tamarind and manna. These were absorbed into Western medicine.

The discovery of the New World then brought others, notably cinchona for malaria: also known as Peruvian or Jesuits' bark, it was the basis of quinine. And in an age in which most remedies were 'simples', that is vegetable derivatives, the rebel Paracelsus was a champion of mineral and metallic remedies – mercury became standard against syphilis – and proclaimed the doctrine of specific remedies for specific diseases. On the model of the bark, Thomas Sydenham (1624–89), the so-called 'English Hippocrates', similarly looked forward to the day when every disease would have its own specific.

From time to time new medicaments were stumbled upon, as with the Rev. Edmund Stone's announcement in the eighteenth century of willow bark as a febrifuge (fever remedy), the first stage on the road to aspirin. Official pharmacopoeias, however, long remained an embarrassing gallimaufry of largely useless remedies, including such relics of magical potions as the bezoar, a stony concretion found in the stomach of ruminants, recommended as a poison antidote. As seen in Chapter 2, Samuel Hahnemann, the founder of homeopathy, was one among many convinced that traditional polypharmacy (a multiplicity of medicines in large quantities) did more harm than good.

It must be remembered, however, that drugs were not expected in humoral medicine to play a decisive role in healing: banking on 'heroic' remedies was what quacks did. Traditional therapeutics had many strings to its bow, including regulation of diet and environment (for instance, travelling for health), and giving wise counsel. A good drug was expected less to zap a disease than, through purging, sweating or cleansing the blood, to aid the healing power of Nature in restoring balance to the system.

25. *Sufferers of syphilis being treated with mercury.*
John Sintelaer, 1709.

In the nineteenth century, however, study of *materia medica* was transformed, slowly and unevenly, into laboratory-based pharmacology, and drugs became production-line items. Initially in France and then in Germany, common plant drugs such as opium were subjected to systematic chemical analysis: the result was the synthesis among others of codeine, nicotine, caffeine, morphine and, later, cocaine. The ability to produce such chemicals in measured, consistent strengths was to prove essential for the mass-production and marketing of medicines.

There was growing symbiosis between drugs research and manufacture, as the booming chemical industry spied profits in pills. Pharmaceutical firms joined hands with academic pharmacology above all in Germany, where major research schools emerged. By 1900 companies were turning lab-made developments to profit – as in the case of aspirin, marketed by Bayer in Germany. In England the Burroughs-Wellcome company funded laboratories to make pharmacology more scientific and pioneer new cures.

The doyen of early twentieth-century researchers was Paul Ehrlich (1854–1915), from 1899 director of the Royal Prussian Institute for Experimental Therapy in Frankfurt-am-Main. Building on bacteriology, Ehrlich had the idea of applying the theory of natural antibodies (resistance agents) to the creation of synthetic drugs. He made a contribution of his own to the immunity debate: the 'side-chain' (or chemical affinity) model. This built on the supposition that an antibody in the blood, produced in response to a certain hostile micro-organism, was specific for that organism and highly effective in killing it, but harmless to the host. Being Nature's remedies, antibodies were thus magic bullets which flew straight to their mark and injured nothing else. Hence the challenge was to find chemical equivalents lethal to a particular organism and non-toxic to its host. Chemotherapy's mission lay in the discovery of synthetic chemical substances which would act exclusively against disease-producing micro-organisms.

Ehrlich addressed that ever-shocking disease syphilis. In 1905 the protozoan parasite causing it, a spiralling thread-like single-celled organism, was isolated from sores and designated the *Spirochaeta pallida* (since renamed *Treponema pallidum*). Diagnostic screening was made possible in 1906 when August von Wasserman developed his famous blood test. Seeking a chemical cure, by 1907 Ehrlich had tried out over 600 arsenical compounds before he took out a patent on Number 606. Within three years about 10,000 syphilitics had been cured with the preparation, by then called Salvarsan.

Would not similar chemical magic bullets rapidly follow for many other diseases? In the event, Ehrlich's hopes were dashed. Hundreds of compounds, including some new synthetic dyes – promising because of their fixing properties – were tried against the common bacterial diseases, but without success. Chemotherapy came to seem, after all, an impossible dream. This situation changed, however, in 1935, thanks to experiments with Prontosil conducted by a fellow German, Gerhard Domagk, director of research of the chemical company Bayer.

Searching, like Ehrlich, for chemical remedies, Domagk found that Prontosil red, a brilliant red dye, cured mice injected with a lethal dose of streptococci. He then successfully treated his daughter for erysipelas, a strep infection. Scientists at the Pasteur Institute in Paris determined that one component in the compound, later called sulphanilamide, was largely responsible for Prontosil's bacteriostatic action – that is, it did not actually kill bacteria, but prevented them from multiplying in the host, thus allowing the body's own immune system to destroy them.

The new drug went into production and, as it could not be patented – Prontosil was basically sulphonamide, which had been synthesized back in 1907 – it became readily and cheaply available. At Queen Charlotte's Maternity Hospital in London, Leonard Colebrook used it to treat that terrible killer, puerperal fever, slashing mortality from 20 per cent to 4.7 per cent and at last realizing Semmelweis's dream (see Chapter 6). He hailed it as a miracle drug.

Though effective against streptococci, sulphanilamide proved little use against pneumococcal infections, which led scientists to look for further 'sulpha' drugs. In 1938 a team at the British manufacturers May and Baker developed M&B 693, which worked well against pneumococci and was even better than sulphanilamide against streptococci. Between them, the new sulpha drugs checked erysipelas, mastoiditis, meningitis and some urinary diseases – sulphanilamide could clear up a case of gonorrhoea in just five days.

Pasteurian bacteriology had meanwhile opened up the prospects of *biological* (as distinct from *chemical*) agents to destroy bacteria. Folk wisdom – for instance, the use of mouldy bread to keep cuts clean – suggested that fungi might be antibacterial, but the first conclusive observation of antibacterial action was made in 1877 by Pasteur himself: while anthrax bacilli multiplied rapidly in sterile urine, he found that the addition of common bacteria halted their development.

The situation in which 'one creature destroys the life of another to preserve its own' was styled 'antibiosis', and the word 'antibiotic' (destructive of life) was later coined by Selman Waksman (1888–1973), a Russian-born soil microbiologist active in the USA. The first such antibiotic was penicillin, a natural by-product from moulds of the genus *Penicillium*, brought to light through the work of Alexander Fleming, a Scottish bacteriologist at St Mary's Hospital, London.

During the First World War, Fleming had been working on resistance to infection, and had concluded that the harsh chemical antiseptics used to cleanse wounds actually damaged the body's natural defences. He was therefore receptive when he discovered in 1922 the anti-bacterial enzyme lysozyme, a component of tears and mucous fluids.

Identification of penicillin came six years after, in August 1928. Fleming had been handling staphylococci, the pathogens responsible for boils, pneumonia and septicaemia. Returning from holiday, he found that a mould which had appeared on a staphylococcus culture left in a petri dish in his laboratory seemed to have destroyed the bacterial colonies. He identified it as *Penicillium rubrum* (actually it was *Penicillium notatum*). While the penicillin destroyed not just staphylococci but also streptococci, gonococci, meningococci and pneumococci – that is, most harmful bacteria – it had no toxic effect on healthy tissues and did not impede leucocytic (white cell) defence functions. However, being hard to produce and highly unstable, it was clinically unpromising; Fleming did nothing and the medical community gave it little notice.

Ten years later, however, a team of Oxford scientists, led by the Australian Howard Florey and including Ernst Chain, a refugee from the Nazis, launched a research project on microbial antagonisms and began to grow *Penicillium notatum*. On 25 May 1940 they inoculated eight mice with fatal streptococci doses, and four were then given penicillin. By next morning all had died except the four treated mice.

Florey seized upon the drug's potential and tried it on human patients, with promising results. He then approached British pharmaceutical companies, but they were too busy supplying wartime needs; so in July 1941 he went to the United States to get the drug rushed into production. In May 1943 he tested it on war wounds in North Africa: the success was phenomenal. Within a year, sufficient was available to allow unlimited treatment of Allied servicemen. Penicillin proved exceptionally effective against pneumococci, gonococci, meningococci and the bacilli of anthrax, tetanus and syphilis. It was the first drug effective against pneumonia. In 1945 Fleming, Florey and Chain shared the Nobel Prize for this wonder drug.

Other antibiotics followed. In 1940 Waksman isolated a fungal antibiotic called actinomycin. Though lethal to bacteria, it proved so toxic that it was not tried clinically, but it convinced him that he was on the right track. In 1944 he hit upon another species of this fungus, from which he isolated the antibiotic streptomycin, which proved active against the tubercle bacillus, while its toxicity was low.

Use of streptomycin rapidly led, however, to the emergence of resistant strains, and it was found more effective against tuberculosis when used in combination with para-amino-salicylic acid (PAS). In 1950 testing began on a third anti-TB agent, isoniazid. Like streptomycin, it too was prone to resistance, but problems were minimized by the combination of these anti-tubercular drugs into a single therapeutic package. The 'white plague' was already in decline but antibiotics delivered the *coup de grâce*.

So long dreamed-of but postponed, the therapeutic revolution was now a reality. New drugs of many kinds followed in the 1950s, notably cortisone, invaluable for rheumatoid arthritis and other inflammatory conditions, and the first effective psychopharmacological agents – lithium, valuable in cases of manic-depression, and chlorpromazine (Largactil) for schizophrenia.

Antibiotics were ineffectual against viruses such as flu, but new anti-viral vaccines began to appear, notably against that scourge of children, polio, the summer crippler. After intense rivalry between Jonas Salk and Albert Sabin as to the preferability of a 'live' or 'dead' vaccine, polio vaccine was introduced in the USA in 1955. A key figure in the fight against viral disease was another American, John Enders, who developed the measles vaccine, licensed in 1963.

Anti-viral drugs proved extremely difficult to develop. Only since the 1970s has progress been made, first with acyclovir, effective against cold sores and herpes zoster (shingles). Many viruses, such as influenza and HIV, continue to outsmart the scientists, since they mutate so rapidly.

If before 1900 the contents of the pharmacopoeia were useful largely, if at all, as placebos, by the 1960s a cornucopia of truly effective drugs had emerged out of the twentieth-century laboratory: antibiotics, anti-hypertensives (beta-blockers) to prevent strokes, anti-coagulants, anti-arrhythmics, anti-histamines, anti-depressants and anti-convulsants, steroids such as cortisone against arthritis, bronchodilators, ulcer cures, endocrine regulators, cytotoxic drugs against cancers, and others besides. The dream of 'a pill for every ill' – Sydenham's 'specifics' – seemed on the way to being realized.

But this golden age was not without disasters. Introduced as a 'safe' sleeping tablet, Thalidomide was withdrawn – belatedly – in 1961 after causing horrendous foetal defects in over 10,000 babies: shockingly, its manufacturer in Germany turned a blind eye to warnings about its appalling side effects. Other tragedies and scandals came to light. From the 1940s the synthetic oestrogen diethylstilbesterol (DES) was given to women to prevent miscarriage. Even after 1971, when it was discovered that DES could cause reproductive problems, including vaginal cancer, in 'DES daughters', it continued to be prescribed in the United States as a morning-after pill. It took such tragedies to bring into being stringent clinical trials for effectiveness and safety and strict licensing procedures. Some claim that the consequent overregulation now discourages bold drugs innovation.

Whatever the reasons, the last few decades have not produced successors comparable to the new miracle drugs of the previous generations: many recent preparations are 'me-too' drugs, minor variants on existing ones, designed to win a rival manufacturer a share of the market. More worryingly, misuse of antibiotics has encouraged drug-resistant strains of tuberculosis and other infections, resurgent from the 1980s, especially in those whose immune systems are AIDS compromised. Drug abuse and dependency – by no means only in the guise of illegal narcotics – looms as an urgent problem for medicine and society alike.

CHAPTER SIX

Surgery

He who wishes to be a surgeon should go to war.
Hippocrates

Surgery is as old as civilization itself, for ancient skull remains show that trepanning (or trephining) was being performed at least as early as 5000 BC. Operators used stone cutting tools to extract portions of the cranium, presumably to relieve sufferers from the torments of 'devils'. Bonesetting was also practised, while Egyptian medical papyri of the second millennium BC describe quite sophisticated surgical procedures for abscesses, minor tumours and disorders of the ear, eye and teeth.

From an early date healers in India were couching for cataract. This involved introducing a thin knife into the eye in front of the lens which had become opaque; this was then pressed backwards into the lower part of the vitreous, where it no longer obscured vision. And Ayurvedic healers even pioneered reconstructive surgery, especially remodelling damaged noses (rhinoplasty). A leaf-shaped flap of skin would be cut from the forehead, making sure that the end nearest the bridge of the nose remained attached.

While primarily about 'physic', the Hippocratic *corpus* includes a wounds treatise. Fractures were to be treated by reduction (restoration of the limb to its normal position) and immobilization with splints and bandages; the surgeon's knife was to be used for excising nasal polyps and ulcerated tonsils; cautery (application of a red-hot iron to sear the flesh) was recommended for haemorrhoids; and trepanning was described. In general, however, Hippocratic wound treatment was limited and conservative. Vascular ligature (tying of veins to arrest blood-flow) was unknown to the Greeks, and internal surgery was avoided – herbal medicines were preferred for cancer, appendicitis, internal stones and so forth.

The Hippocratic Oath directed physicians to leave knifework strictly to the surgeon. While recognizing the surgeon's skills, this bred an enduring medical division of labour in which surgery was viewed as inferior, the work of hand not head. Certain Ancient physicians, however, paid attention to matters surgical. Soranos of Ephesos (AD 96–138) wrote extensively on obstetrics, discussing the use of the birthing-chair and giving instructions for difficult birth positions. He described, for instance, the procedure which was later called 'turning the foot' (podalic version), easing a hand into the womb and pulling down a leg, so that the baby would be born feet-first.

Islamic medicine set greater store by surgery, perfecting cautery with an iron to staunch bleeding. In his great *Altasrif* (Collections), Albucasis, practising in tenth-century Spain, discussed a multitude of operations, but placed greatest faith in cautery. Meanwhile, among medieval Christians, the Salernitan school, flourishing in southern Italy from the eleventh century, explained surgical handicraft.

Wound management grew controversial. Hippocratic medicine had held that suppuration was indispensable for healing, since pus derived from poisoned blood, which needed to be expelled – a view which provided authority for the long influential doctrine of 'laudable pus': pus was beneficial and its formation should be encouraged. The counter idea of dry (pus-free) wound management was advanced in distinguished treatises by the Frenchmen Henri de Mondeville (b. 1260) and Gui de Chauliac, whose *Grande Chirurgie* (1363) was for two

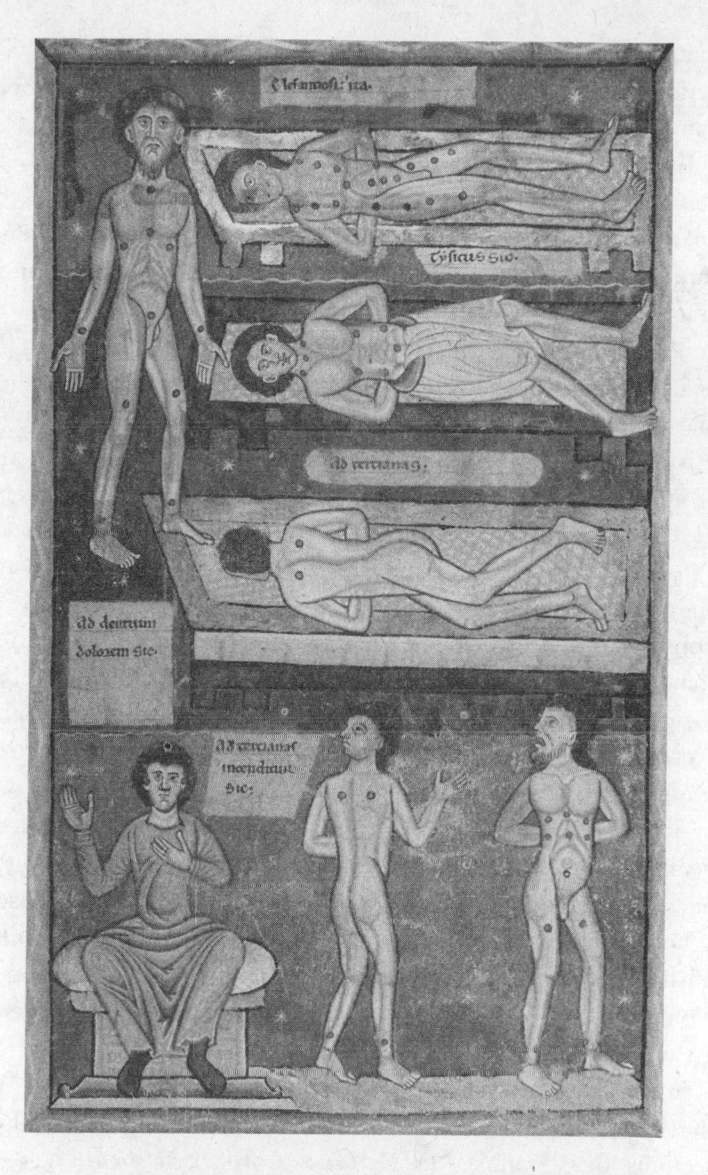

26. Chart showing cautery points on various parts of the body, 1462.

centuries the prime surgical text. Gui held that wounds healed better without suppuration. As with anatomy, it took a bold man to challenge the authority of the Greeks.

Gangrenous wounds evidently required amputation, though before the sixteenth century it was rarely performed above the knee: patients would bleed to death. Experience taught medieval surgeons to remove more bone while preserving the maximum amount of soft tissue, thus permitting skin to mend over the bone and in due course form a usable stump, to which a peg leg or hook might be attached. Cautery with a hot iron or boiling oil remained the main means of checking haemorrhage.

Many surgeons learned or developed the cutter's art in the army – the battlefield was proverbially the school for surgery. The introduction of gunpowder in the late Middle Ages exacerbated the character of wounds. Lead bullets tore through flesh and shattered bones, driving foreign matter deep, and so infections became a critical problem, giving rise to the belief that some kind of gunpowder poison had entered the wound.

In northern Europe civilian surgery was performed by operators who doubled as barbers (they used the same tools of the trade). It was also undertaken by itinerants (quacks) specializing in one particular operation: there were travelling tooth-drawers, oculists who would couch for cataract, lithotomists who removed bladder stones, and 'hernia masters', who fitted trusses. Whoever undertook it, operative surgery was a risky and painful business; it required 'an eagle's eye, a lion's courage and a woman's hand' – and (perhaps most important for the patient) great speed.

From the sixteenth century, however, surgery was growing more systematic. Ambroise Paré, a towering figure, had sections of Vesalius's *De Fabrica Corporis Humani* (1543) translated into French as part of his *Anatomie Universelle du Corps Humain* (1561), to make the new anatomical teachings available to barber-surgeons lacking higher education. Born in 1510 in northern France, Paré was apprenticed to a barber-surgeon and then saw extensive military service. Developed in the light of battlefield experience, his innovations included vascular ligature (vital for amputations) and a replacement for hot-oil cautery to cleanse wounds. As related in his *Method of Treating Wounds* (1545), he concocted an ointment (or 'digestive') from egg yolk, rose oil and turpentine, which he applied to open or bleeding wounds. The mixture proved successful, and he abandoned the excruciating hot-oil treatment.

In England, John Woodall's *The Surgeon's Mate* (1617) did long service as a manual of naval surgery, as did Richard Wiseman's *Several Chirurgical Treatises* (1676). Wiseman, the 'father of English surgery', picked up much of his experience during the English Civil War, and his account of military surgery reveals its horrors: cannonballs and gunshot caused horrifying wounds, and amputation and trepanation were often the only remedies, performed on the battlefield or on a storm-tossed vessel.

Alongside routine procedures the art spawned a weird-and-wonderful penumbra of unfulfilled promises. There was much to do in the seventeenth century, for instance, about the 'wound salve' announced by the gentleman-scholar Sir Kenelm Digby. Touted to heal rapier wounds, this was a mixture of earth-worms, iron oxide, pig's brains, powdered 'mummy' (embalmed corpse) and other exotic ingredients. The salve was applied not to the wound but to the offending weapon, and was said to work by sympathetic magic. The shortcomings of regular surgery explain the appeal of such pipedreams.

Before the introduction of anaesthesia in the 1840s, invasive surgery was limited in scope; lengthy operations, or ones demanding great precision, were out of the question. A brave man – Samuel Pepys was one – might risk having a bladder

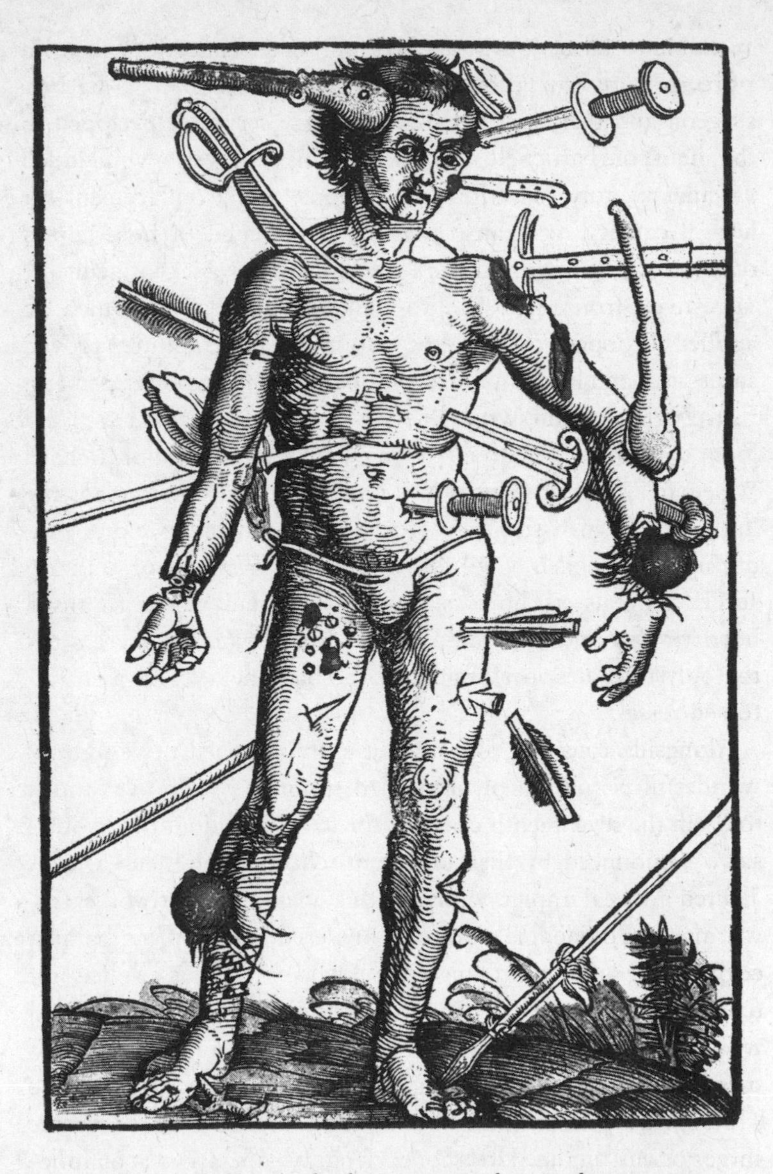

27. *First-aid chart for barber surgeons showing how to treat wounded soldiers. Woodcut, sixteenth century.*

stone removed surgically: luckily he survived, and he had the offending stone mounted as a trophy, writing in his diary on 26 March 1660: 'This day it is two years since it pleased God that I was cut of the stone at Mrs Turner's in Salisbury Court, and did resolve while I live to keep it a festival.' And a few highly dangerous operations were performed in dire emergency, including caesarean section. There is no record till the 1790s of a caesarian in Britain which the mother survived.

Everyday surgical work necessarily remained small-scale and fairly safe, if often pretty painful: dressing wounds, drawing teeth, treating syphilitic chancres and sores (all too common from the sixteenth century), lancing boils, trussing-up ruptures, and so forth. The most frequent procedure – the surgeon's bread and butter – was blood-letting. It followed from humoral doctrines, especially Galen's theory of 'plethora' – the idea that fevers, apoplexy and headache resulted from an excessive build-up of blood. Cupping with scarifications was another much used procedure for drawing blood or boils.

Snobbishly disparaged as a manual skill rather than a liberal science, this 'cutter's art' was traditionally subordinate to physic in the medical pecking order. Organized into trade guilds, surgeons had normally passed not through an academic but a practical education, via apprenticeship. They carried low prestige – the bloody and blundering 'Mr Sawbones' is a standard butt in plays and prints. But from the eighteenth century surgery began a long and lasting rise.

Among practical improvements, a superior method for removing bladder stones, lateral cystotomy, was introduced around 1700 by the itinerant practitioner who styled himself Frère Jacques – he wore a Franciscan friar's habit to ensure safety on his travels. He is credited with some 4,500 such lithotomies as well as 2,000 hernia operations. Johannes Rau in Amsterdam and William Cheselden in London took up his

28. *An ill man being bled by his doctor. Etching, James Gillray, 1804.*

method. The latter won fame for excising bladder stones with exceptional rapidity – whereas other surgeons might take twenty, he could complete the excruciatingly painful knife-work, performed without anaesthetics, in a couple of minutes flat. As a result he commanded huge fees, apparently up to 500 guineas, and equal respect: 'I'd do what Mead and Cheselden advise/To keep these limbs and to preserve these eyes,' sang Alexander Pope.

Other operations also underwent refinement. The celebrated French military surgeon Jean-Louis Petit developed new practices with amputations at the thigh, thanks to the use of an effective tourniquet which controlled blood-flow, used in combination with the vascular ligatures as advocated by Paré.

Military surgery advanced, particularly the management of gunshot wounds. By the early eighteenth century the British fleet had 247 vessels, each carrying a surgeon and his mate. For those with strong stomachs, such as the surgeon-hero of Tobias Smollett's 1748 novel *Roderick Random*, naval or military service provided invaluable experience and a professional *entrée*.

Not least, obstetrical skills were progressing. Childbirth had traditionally been exclusive to women: the mother, her female kith and kin and a midwife, who made up in experience what she lacked in formal training. Initially among polite society in England and later in North America, this traditional 'granny midwife' figure became displaced, however, by a male surgeon, the 'man-midwife' or *accoucheur*. He claimed superior skills: by dint of being a qualified medical practitioner, armed perhaps with an Edinburgh degree, his anatomical expertise made him confident that he could leave normal deliveries to Nature, while teaching him how to handle emergencies.

Contrary to common modern opinion, such leading obstetricians as William Hunter, a Scot who attended Queen Charlotte,

29. *'The man-mid-wife'. A full frontal picture divided in half, one side representing a man, the other a woman. Etching, I. Cruikshank, 1793.*

George III's wife, prided themselves upon being less interventionist than the midwives they displaced. Yet *accoucheurs* also possessed, unlike the midwife, surgical instruments, above all the new obstetric forceps, for use in difficult labours and emergencies. Introduced in the seventeenth century and initially kept secret by their inventors, the Chamberlen family, forceps had become a familiar tool of the trade by 1730.

In America midwifery became medicalized under the inspiration of William Shippen, who had studied at Edinburgh and with the Hunter brothers in London. He taught anatomy and midwifery in Philadelphia from 1763, helping to establish the male domination of obstetrics which became so marked in the USA.

Where *accoucheurs* flourished, childbirth was transformed, and baby-rearing with it. A fashionable lady of the late eighteenth century might now opt to have her husband present at labour, giving birth in a room into which daylight and fresh air were admitted. Her newborn would no longer be swaddled: enlightened thinking taught that freedom for infant limbs would strengthen bones and promote healthy development. On medical advice, the modern mother *à la mode* also now breast-fed: mother's milk was surely best and would encourage mother–baby bonding. Progressive surgeons – the Dr Spocks of their day – thus played a part in changing the theory and practice of childbirth and baby-care.

Boosted by such improvements, surgery rose in professional standing, first of all in France. As elsewhere, French practitioners were initially barber-surgeons, but they succeeded in emancipating themselves from their lesser half. The breakthrough came in 1731, when a royal charter established the Académie Royale de Chirurgie; in London the Company of Surgeons split from the barbers a few years later in 1745, the

30. A woman giving birth aided by a male surgeon, who fumbles beneath a sheet to save the woman from embarrassment. Wood engraving, 1711.

first step in the transformation of the Company into a College.

In France the tradition of training surgeons by apprenticeship came to an end in 1768; and thereafter French surgeons vied with physicians in status, claiming that surgery was no mere manual art but a science. The relocation of surgical education into the hospital reinforced the links, growing since Vesalius, between surgery and anatomy, and pointed towards the patho-anatomical perspective on disease dominant in Revolutionary Paris (see Chapters 3 and 4). Thanks to these developments, France led the way in surgery, drawing in students from all over Europe.

Parallel changes were occurring elsewhere, however. It is significant that Alexander Monro, the first professor of anatomy in the new Edinburgh medical school, founded in 1726, was himself a surgeon (he was followed in the chair by his son and grandson, also Alexanders!). The quality of the combined medical and surgical education given north of the Border began to blur the old distinctions between the professions.

New private anatomy schools in London further elevated surgery's prestige. Among the most illustrious, William Hunter's in Piccadilly offered instruction in anatomy, surgery, physiology, pathology, midwifery and the diseases of women and children. Addressing such key surgical topics as inflammation, shock, disorders of the vascular system and venereal disease, his younger brother John's four main treatises – *Natural History of the Human Teeth* (1771), *On Venereal Disease* (1786), *Observations on Certain Parts of the Animal Oeconomy* (1786) and *Treatise on the Blood Inflammation and Gunshot Wounds* (1794) – were hailed as raising surgery from craft to science, thanks to his grasp of physiology.

The success of the Edinburgh University medical school and of private anatomy schools spawned one nagging problem: the shortage of bodies legally available for dissection (see Chapter 3).

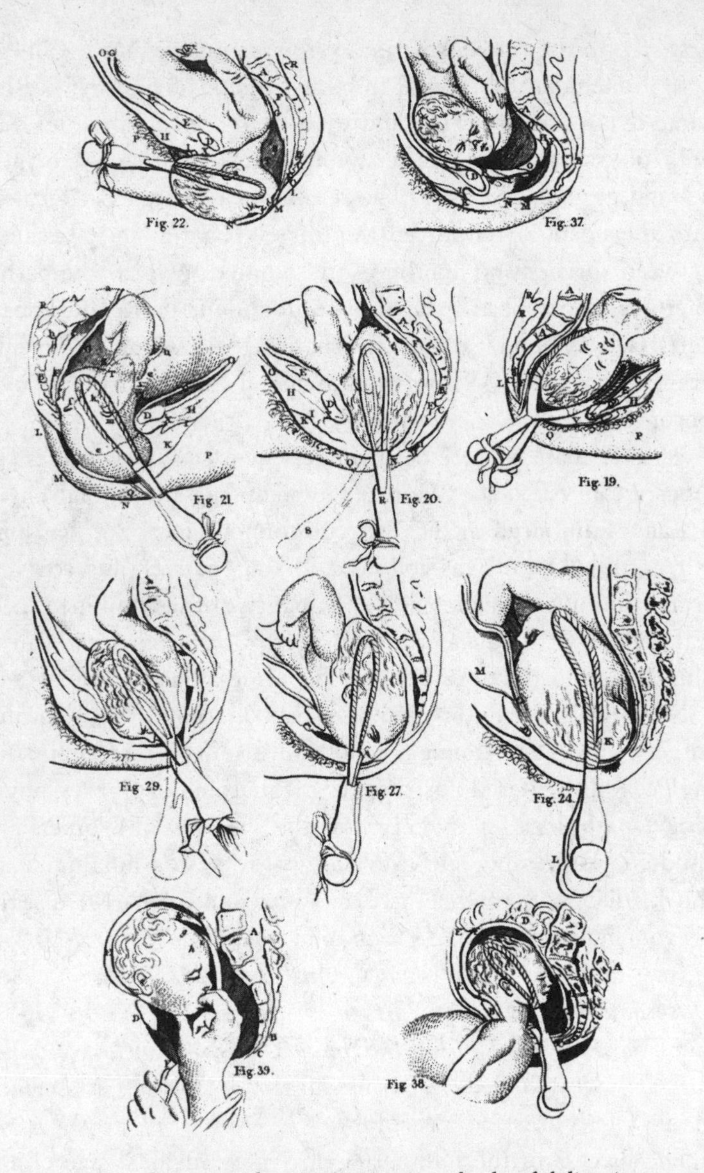

31. Ten diagrams showing various methods of delivering a baby using forceps. Etching, 1791.

The quick way to overcome this was by resort to illicit grave-robbers, the 'sack-'em-up men', who supplied the anatomists (who asked no questions). The two most notorious such 'resurrectionists', William Burke and William Hare in early nineteenth-century Edinburgh, took an even shorter cut: they murdered their victims, before selling them for research.

The first half of the nineteenth century brought a few daring new operations, especially in the New World. In 1809 the American surgeon Ephraim McDowell performed the first successful ovariotomy, without anaesthesia, on the 47-year-old Jane Todd Crawford, removing fifteen pounds of a 'dirty gelatinous substance' from her ovarian cyst. Remarkably, she lived a further thirty-one years. Another American, John Attlee, removed the ovaries of seventy-eight women between 1843 and 1883, with sixty-four recoveries. Overall, however, operative surgery's scope remained restricted and its success uncertain, before two critical innovations: anaesthesia and antisepsis.

Medicine had always made some use of analgesics, and from early times the pain-deadening qualities of opium, hashish and alcohol were familiar. The first gas known to have anaesthetic powers was nitrous oxide, the object of self-experimentation in the 1790s by the Bristol physician Thomas Beddoes and his brilliant young assistant Humphry Davy. Yet operations were traditionally performed on patients who were conscious. As the account by the novelist Fanny Burney of the removal of her cancerous breast makes amply clear, the knifework involved was inconceivably agonizing.

The real breakthrough in practical anaesthetics came in January 1842 when William E. Clarke, a practitioner from Rochester, New York, extracted a tooth under ether. Its use spread to Europe. On 21 December 1846 Robert Liston, a top London surgeon renowned for his speed, amputated the diseased thigh of a patient unconscious under the vapour, going on to sing the praises of this 'Yankee dodge'. Ether was quickly replaced, however, by the safer chloroform. On 19 January 1847 James Young Simpson of Edinburgh used it for the first time to allay labour pains. Chloroform anaesthesia soon became widespread for childbirth – resistance from those who quoted the biblical pronouncement that women should bring forth in pain was stilled when Queen Victoria was chloroformed for the birth of Prince Leopold on 7 April 1853.

The introduction of effective anaesthesia made otherwise unbearably traumatic internal operations feasible. 'That beautiful dream has become a reality: operations can now be performed painlessly,' declared the distinguished German surgeon Johann Dieffenbach, on seeing his first anaesthetized patient undergo an operation. But because of the appalling post-operative death-rate of invasive surgery, due to septicaemia, anaesthesia did not by itself revolutionize practice, for the menace of infection was unremitting. Working in 1848 in the maternity wards of the Vienna General Hospital, Ignaz Semmelweis was appalled by the dreadful fatality levels from puerperal fever. The first obstetrical clinic, run by medical men, had, he observed, a much higher death rate than the second, run by midwives. Why? It was because, he deduced, medical staff and students went directly from the post-mortem to the delivery rooms, thereby spreading infection. He instituted the rule of washing hands and instruments in chlorinated lime solution between autopsy work and handling patients, and the mortality rate dropped to the same level as in the second clinic.

Opposition to his shocking views – doctors spread infection! – led him to quit Vienna in 1850, and, resentful and frustrated, he was to die in a lunatic asylum. Antagonism to Semmelweis was no blatant professional closing of ranks, however, but was consistent with the aetiological theories of the time. Infections, as we have seen, were thought to be caused by miasmata exuded by the soil and other non-human sources. Adherents of such views – they included Florence Nightingale – gave priority as preventive measures to ventilation and the avoidance of overcrowding in hospitals.

Antiseptics – that is, substances or procedures designed to counter putrefaction or infection – were far from unknown. Wine and vinegar had long been used for treating wounds, and around 1820 iodine became popular. It was, however, Joseph Lister who first developed effective antiseptic techniques and campaigned tirelessly on their behalf.

Born into a Quaker family, Lister graduated from London University and rose to become regius professor of surgery in Glasgow. In 1861 he was put in charge of the new surgical block in the Royal Infirmary, where he evolved his methods. Believing that carbolic acid (phenol) would be effective as an antiseptic, he undertook his first trial on 12 August 1865, on an eleven-year-old boy, James Greenlees, whose left leg had been run over by a cart. He dressed a compound fracture of the tibia with lint soaked in linseed oil and carbolic acid, and kept the dressing in place four days. The wound healed perfectly and a healthy James walked out of the Infirmary six weeks later.

Publicizing his methods in *Lancet* in 1867, Lister insisted on two points: germs caused infections; and – for all the old ideas about 'laudable pus' – infection and pus-formation were not inevitable, still less beneficial, stages in wound healing.

An early opportunity for putting Listerian practices to the test came in the Franco–Prussian War (1870–71), when the German military medical staff introduced some of his procedures in treating battle wounds. They achieved superior outcomes to the French, who neglected Lister completely.

By 1890 antiseptic surgery had established itself – and Lister's messy and smelly carbolic spray was fast being replaced by less harsh antiseptics. Heat sterilization of instruments was urged by Koch in 1881, and the American surgeon William S. Halsted, of the Johns Hopkins Hospital, introduced the use of rubber gloves. By 1900 operations were no longer an unedifying spectacle of surgeons, clad in blood-caked frock-coats, wielding the knife in dingy rooms with sawdust-covered floors. Face-masks, rubber gloves and surgical gowns lessened the risks of infection, and sterile environments had become *de rigueur*. The modern spotless and gleaming operating theatre was emerging and success rates rising.

As late as 1874, a leading English surgeon could opine that 'the abdomen, chest and brain will forever be closed to operations by a wise and humane surgeon'; and Lister himself rarely probed into major cavities, confining himself mainly to setting fractures. But things were changing: thanks to anaesthetics and antiseptics, surgery's horizons opened dramatically. In Vienna the celebrated Theodor Billroth (1829–94) made important innovations, pioneering abdominal surgery and cutting for various cancers, especially of the breast. In America, Halsted devised radical mastectomy, which remained for many years the favoured treatment for breast cancer. Appendectomy was developed: in 1902 Edward VII was operated on when his appendix erupted just before his coronation. Just as Queen Victoria played her part in the acceptance of anaesthesia, her son had a role in modern surgery. Cholecystectomy, excision of the gall bladder, was introduced in 1882, and removal of gallstones became routine. Surgery on the small intestine, notably for cancer, was begun around the same time, as were prostate operations. Surgery was also introduced for traditional medical conditions, such as tuberculosis. Pneumothorax (surgical collapsing of the lung so as to rest it) enjoyed a brief vogue.

Two surgeons were even honoured at this time with a Nobel

Prize – Theodor Kocher in 1909 for his work (partly surgical) on the thyroid gland, and Alexis Carrel in 1911 for his studies of tissue culture and techniques of suturing blood vessels. Carrel's fine needle-work paved the way for surgery on aneurysms, varicose veins and blood clots, and the use of replacement tissues. The problem of rejection encountered later proved a major obstacle to transplants.

As the twentieth century unfolded, surgery seemed to know no limits. Such progress would have been quite impossible without key technological innovations which enabled the interior of the body to be imaged and monitored. A massive advance was Wilhelm Röntgen's discovery of X-rays in 1895. Around 1900 Willem Einthoven of Holland devised the first electrocardiograph, which picked up the electrical activity of the heart, thus making possible effective monitoring of cardiac disorders. Catheterisms in turn permitted investigation of heart and liver functions. From the mid 1950s ultrasound, developed in Sweden and the USA, proved surgically valuable in cardiac diagnosis and for assessing foetal progress through pregnancy. Visual diagnostics were further boosted with the devising in 1972 of the computerized tomograph (CAT-scan) by Godfrey Hounsfield, alongside PET (positron-emission tomographic scanning) and magnetic resonance imaging (MRI) – the last was capable of showing metabolic organs by using radio waves.

Flexible endoscopes, drawing on glass-fibre optics, were used from the 1970s, first for diagnostics but soon also for therapeutic interventions, not least in connection with lasers, those 'optical knives' which have proved so valuable in eye as well as in internal surgery. Today, thanks to telescopic microscopes, keyhole surgery has become common for hernias, the gall bladder, the kidney and the knee-joint.

Guided, from X-rays onwards, by such aids to seeing into

32. Surgeons examining a man's chest using an X-ray without any protective clothing. W. Small, 1900.

inner space, surgeons grew ever-more ambitious. Initially the modern surgeon's attention was mainly directed to tumours and infections which caused obstruction or stenosis (constriction of vessels), above all in the digestive, respiratory and urogenital tracts. These could be relieved by cutting or excision. All the cavities and organs of the body yielded to the all-conquering knife: the abdomen, the thorax, even the cranium. Indeed, the hope that 'a chance to cut is a chance to cure' led to psychosurgery, in the form of lobotomy and leucotomy: by 1951 over 20,000 patients in the USA had undergone these rash if well-meant procedures. For centuries ultra-cautious, surgery acquired a cavalier air. The Ulsterman Sir William Arbuthnot Lane advocated removal of yards of the gut, sometimes for ordinary constipation, or even as a prophylactic measure. Much other unnecessary and even dangerous surgery was ventured. Between 1920 and 1950 hundreds of thousands of tonsillectomies were performed, almost all quite needless, while hysterectomies enjoyed a similar fad. Compare the vogue for caesarians nowadays.

Surgery's rapid advance was given further momentum by external events, especially war and traffic accidents. The use of high explosive shells made battle injuries more ghastly than ever. One response to such wounds and burns was the pioneering of plastic and reconstructive surgery, particularly for the face, notably by Sir Harold Gillies in the First World War and his cousin, Archibald Hector McIndoe, during the Second. Essential to crisis surgery, the establishment of blood and plasma banks was also sped by war. During the Spanish Civil War techniques were developed for administering stored blood by indirect transfusion into the patient from a bottle. First tried, briefly, in the seventeenth century, blood transfusions, essential to the modern operating theatre, had finally been made safe and effective.

By the second half of the century antibiotics and better immunological knowledge were further extending the scope of operability. Surgery could as a result be performed on cases hitherto deemed too risky because of the danger of infection, for instance, interventions in the lung in contact with atmospheric micro-organisms.

The development of such capacities to manage cardiac, respiratory and kidney function and fluid balance ushered in a new phase: the transition from *removal* to *restoration* and *replacement* surgery. Implants are a good marker. The first implantation of an artificial apparatus came in 1959 with the heart pacemaker, developed in Sweden by Rune Elmqvist. Implants nowadays include eye lenses, cochlear implants, vascular protheses and heart valves, while artificial protheses such as metal-and-plastic hip joints (introduced in 1961) have become routine. Not all such implants have been for health reasons, however; witness the boom in silicon breast implants alongside other forms of cosmetic surgery – over 800,000 'facials' are annually performed in the United States alone.

This transition to restoration and replacement is conspicuous in cardiac surgery. The heart had always been a no-go area. An early development in the 1920s was cutting for constriction of the mitral valve – the valve between the left auricle and ventricle – which results in impaired blood circulation. Hopes followed that congenital heart disease (the blue-baby syndrome) could be rectified surgically. Such hole-in-the-heart babies were blue because of inborn anomalies which meant that blood was passing directly from the right chamber of the heart to the left without being oxygenated in the lungs. The operation was first undertaken at the Johns Hopkins Hospital in 1944.

The most dramatic advances, however, were made possible by the heart/lung machine, designed to bypass the heart and maintain circulation artificially while surgery was conducted on the stopped heart. Such open-heart surgery, enabling surgeons to replace diseased valves or repair defects in the walls between the chambers, began in 1952 in the USA with the implantation of valvular protheses. Within a couple of decades cardiac bypasses had become common, and heart surgery quite routine. These days 200,000 heart procedures a year are performed in the USA.

The most spectacular example of replacement surgery has been transplantation. Successful skin grafts had been performed by the Swiss J. L. Reverdin as early as the 1860s, and these it was which paved the way for reconstructive surgery undertaken by Gillies on First World War casualties. But grafts and transplants had to overcome the problem of rejection – the body's natural response to an invader is to repulse it. The major contributions came from the immunologist Macfarlane Burnet and the biologist Peter Medawar. Around 1960 the first effective immuno-suppressant drugs were introduced. By blocking the production of antibodies without producing life-threatening susceptibility to infections, such drugs, above all cyclosporine in the 1970s, made organ replacement viable.

Kidney transplants came first in 1963, but world news was made in 1967 when Christiaan Barnard at the Groote Schuur Hospital, Cape Town, sewed a woman's heart into Louis Washkansky, who lived for eighteen days. A second patient, Philip Blaiberg, survived for over a year and a half. Early difficulties were overcome, and heart transplants became routine: by the mid 1980s 2,000 were being conducted each year in the USA alone, with two thirds of the recipients surviving for five years or more. Liver and lung transplantation, performed since the 1960s, also became common, as has multiple-organ transplantation.

Life-savers though they are, organ transplants raise acute ethical and legal predicaments. Under what circumstances might a living person ethically become an organ donor? Should there be a for-profit market in the organs of the dead – organs for sale? Should the dead automatically be assumed to consent to removal of organs? And at what point is a person – particularly one kept alive on a ventilator – truly dead, thus authorizing organ removal? Public suspicions about body cannibalization and other questionable practices have led to renewed fears of body-snatching and to a marked reluctance to signal willingness to be a donor.

Equally profound moral and social problems attend advances in reproductive technology made possible since the first test-tube baby in 1978. Together with Robert Edwards of Cambridge University, the surgeon Patrick Steptoe worked on *in vitro* fertilization of human embryos, which resulted in the birth of Louise Brown through IVF and implantation in her mother's uterus. The further practice of surrogate motherhood has enabled otherwise infertile couples to have children of their own, but has raised heated debate, centring on ownership of the embryo. Such ethical questions continue to proliferate, for instance after heavy hormonal treatment and *in vitro* fertilization allowed a 61-year-old Italian woman to give birth in 1992. Similar ethical issues surround sexual surgery performed to change a person's gender.

Surgery has been revolutionized in the last century and a half. For millennia so limited, the craft has become one which knows no frontiers. The new heroic excision techniques of a century ago in turn gave way to the age of restoration and replacement. The more systemic approach to treatment required by replacement therapy is nowadays challenging, perhaps dissolving, the ancient professional boundary between surgery and other medical disciplines and illustrating medicine's increasingly interdisciplinary character. In the process, the surgeon, in the past so

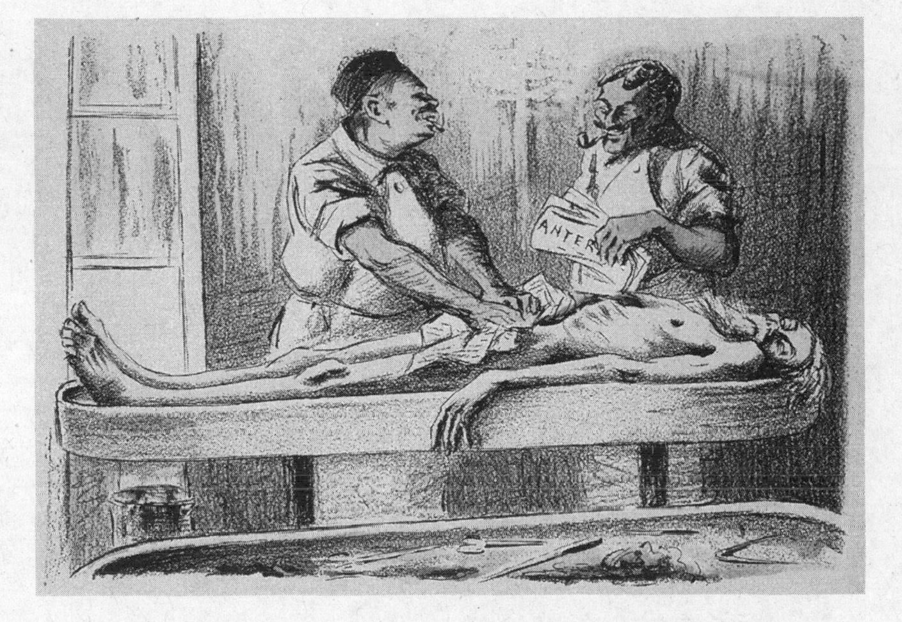

33. A corpse is stuffed with newspapers after being raided for organs by two surgeons. Lithograph, N. Dorville, 1901.

often rather despised, became the superstar of modern medicine. The twenty-first century will rapidly go beyond replacement, deep into the realms of transformative and other elective surgery.

CHAPTER SEVEN

The Hospital

> It may seem a strange principle to enunciate as the very first requirement in a Hospital that it should do no harm.
>
> Florence Nightingale

Today's hospital is to medicine as the cathedral to religion and the palace to monarchy. It is the heart of the enterprise, the site where medicine is practised at its most advanced, specialized, innovative, complex – and costly! In the developed world, hospitals claim the lion's share of the medical budget. And they are the institutions over which medical politics as well as economics are fought: hospitals are always in the news.

But if the high-tech hospital is the jewel in the crown, it was not always so. Medicine initially made do entirely without hospitals; and for long they were marginal – indeed, many people were sceptical about their value.

Classical Greece had no hospitals. The sick might, as noted in Chapter 2, visit a healing shrine, but such religious healing was discounted by the new secular style of medicine promoted by Hippocratic physicians. Imperial Rome for its part provided some hospital facilities, but only for slaves and soldiers. It was with the Christian era that institutions began to be dedicated to the treatment of the civilian sick.

That is no accident, for holiness and healing went together. Christ had performed healing miracles, giving sight to the blind and making the lame walk, and charity was the supreme Christian virtue – witness the parable of the Good Samaritan.

As expressions of Christian charity, compassion and care, the ideals of nursing and healing gave impetus to hospital foundation. After the conversion of the Emperor Constantine early in the fourth century, hospitals sprang up as pious foundations, generally in connection with religious orders devoted to the service of God and man.

During the medieval centuries, thousands were established through pious bequests under the aegis of monks, nuns and others in religious orders. Such hospitals were often short-lived, and they were typically modest, possessing perhaps a dozen beds and a couple of brethren in charge and being organized around the religious offices. It was more important to ensure that Christians died in a state of grace, having made confession and received the sacraments, than to attempt heroic medical treatments. While they sheltered the sick and needy, hospitals were not, in general, centres of specialized medicine: they were more like hospices, that is, places providing refuge and care.

In big cities hospitals became conspicuous fixtures. By the seventh century some hospitals in Constantinople (then capital of what remained of the Roman Empire) had separate wards for men and women, and special rooms for surgical and eye cases. Islam held similar views on pious charity and, from the tenth century, there were multi-functioned hospitals ('bimaristans') in Cairo, Baghdad, Damascus and other Muslim cities. Some of these became used for medical teaching.

To contain a frightening disease, specialized leper asylums were built, where the 'unclean' might be forcibly confined. By 1225 there were up to 19,000 such leprosaria in Europe. As leprosy declined, these were requisitioned for persons suspected of carrying infectious diseases, the insane and even the indigent. When bubonic plague struck in the fourteenth century, leprosaria were also turned into the first plague hospitals. Quarantine lazarettos (named after the protective patron, Saint

Lazarus) began to be deployed to safeguard trade and to protect city populations. The first such pesthouse was built at Ragusa (modern Dubrovnik) in 1377, while Venice enforced quarantine in lazarettos from 1423.

In Venice, Bologna, Florence, Naples, Rome and other major Italian cities, hospitals were to assume a key role in caring for the poor, old and unwell. By the fifteenth century there were thirty-three in Florence alone – one per 1,000 inhabitants. Seven were principally dedicated to the sick, with designated medical staffs. In London St Bartholomew's dates from 1123 and St Thomas's from around 1215. By the close of the fourteenth century there were nearly 500 hospitals in England, though outside the capital and a few other cities they were generally tiny.

The dissolution of the monasteries and chantries during the Henrician and Edwardian Reformations (1536–53) brought the closure of practically all such foundations, as the Crown seized their land and wealth. A handful were re-established, however, on a new and secular basis, including St Bartholomew's and St Thomas's, and also Bethlem (Bedlam), England's only lunatic asylum. Beyond London there were no medical hospitals at all in Britain as late as 1700.

In Catholic countries, and in Protestant Germany, no equivalent of the Henrician asset-stripping occurred, and in Renaissance Spain, France and Italy foundations continued to grow in numbers, size, wealth and power. The Hôtel Dieu in Paris was a huge healing institution, run, right up to the French Revolution, by religious orders. Throughout France, the *hôpital général* (similar to the English poorhouse) emerged in the seventeenth century as an institution designed to shelter, and confine, beggars, orphans, vagabonds, prostitutes and thieves, alongside the sick and mad poor. Basic medical needs were met.

Hospital building could become a prestige project. The gem among continental hospitals was Vienna's 2,000-bed Allgemeine

Krankenhaus (general hospital), rebuilt by the Emperor Joseph II in 1784, a conspicuous expression of the drive of enlightened absolutist rulers towards administrative centralization. Similar in its goals, Berlin's Charité was rebuilt in 1768 by Frederick the Great, while in St Petersburg Catherine the Great erected the huge Obuchov Hospital.

To fill a yawning gap, new hospitals for the deserving poor were founded in eighteenth-century Britain. Crown and Parliament played no part – the organizing zeal and funds came from the charitable impulses of the affluent public at large. The capital benefited earliest. To the metropolis's two medieval foundations, a further five general hospitals were added: the Westminster (1720), Guy's (1724), St George's (1733), the London (1740) and the Middlesex (1745). By 1800 London's hospitals were handling over 20,000 patients a year.

The Edinburgh Royal Infirmary was set up in 1729, followed by hospitals in Winchester and Bristol (1737), York (1740), Exeter (1741), Bath (1742), Northampton (1743) and some twenty other provincial cities. By 1800 every sizeable town had its hospital: England had caught up with the rest of western Europe.

Similar developments took place, though somewhat later, in North America. The first general hospital was founded in Philadelphia in 1751; some twenty years later the New York Hospital was established, and the Massachusetts General followed in 1811, catering for the sick poor. By the early twentieth century America possessed over 4,000 hospitals, and few towns were without one.

Complementing general hospitals, specialist institutions were also founded. London's Lock Hospital, exclusively for venereal cases, opened in 1746. Another novel institution was the lying-in or maternity hospital. The first ones in London were erected around the mid eighteenth century. Some took in unmarried mothers and offered teaching and practice to medical students.

Another development gathering momentum from the eighteenth century was the madhouse, later known as lunatic asylum or mental hospital. Most nations developed a mixed economy of asylums public and private, religious and secular, charitable and for-profit. The more enlightened were expressions of the psychiatric conviction that removal to a well-designed institution was positively therapeutic, though some always functioned merely as convenient places for shutting up inconvenient people. As legally enforced certification procedures evolved in the nineteenth century, such asylums grew ever larger, and silted up with hopeless cases. Before the de-institutionalization movement of the 1960s, around half a million people were locked up in psychiatric hospitals in the USA, and some 150,000 in the UK.

Pre-modern hospitals were very different from today's. While they provided treatment, food, shelter and a chance for convalescence, with rare exceptions general hospitals were not centres of advanced medicine. Most restricted themselves to accidents and casualties and to fairly routine complaints likely to respond to rest and treatment – for example, winter bronchitis or ulcerated legs. Infectious cases were excluded as nothing useful could have been gained by allowing fevers inside: they could not be cured and were sure to spread like wildfire.

Indeed, the fact that hospitals even so became riddled with infections called the very institutions into question: did they not spread the maladies they were meant to relieve? We have seen in the previous chapter Semmelweis's exposure of the lethal maternity wards of the Vienna Allgemeine Krankenhaus. Regarding

34. (overleaf) *Interior of a ward at the Middlesex hospital. Aquatint, J. Stadler, 1808.*

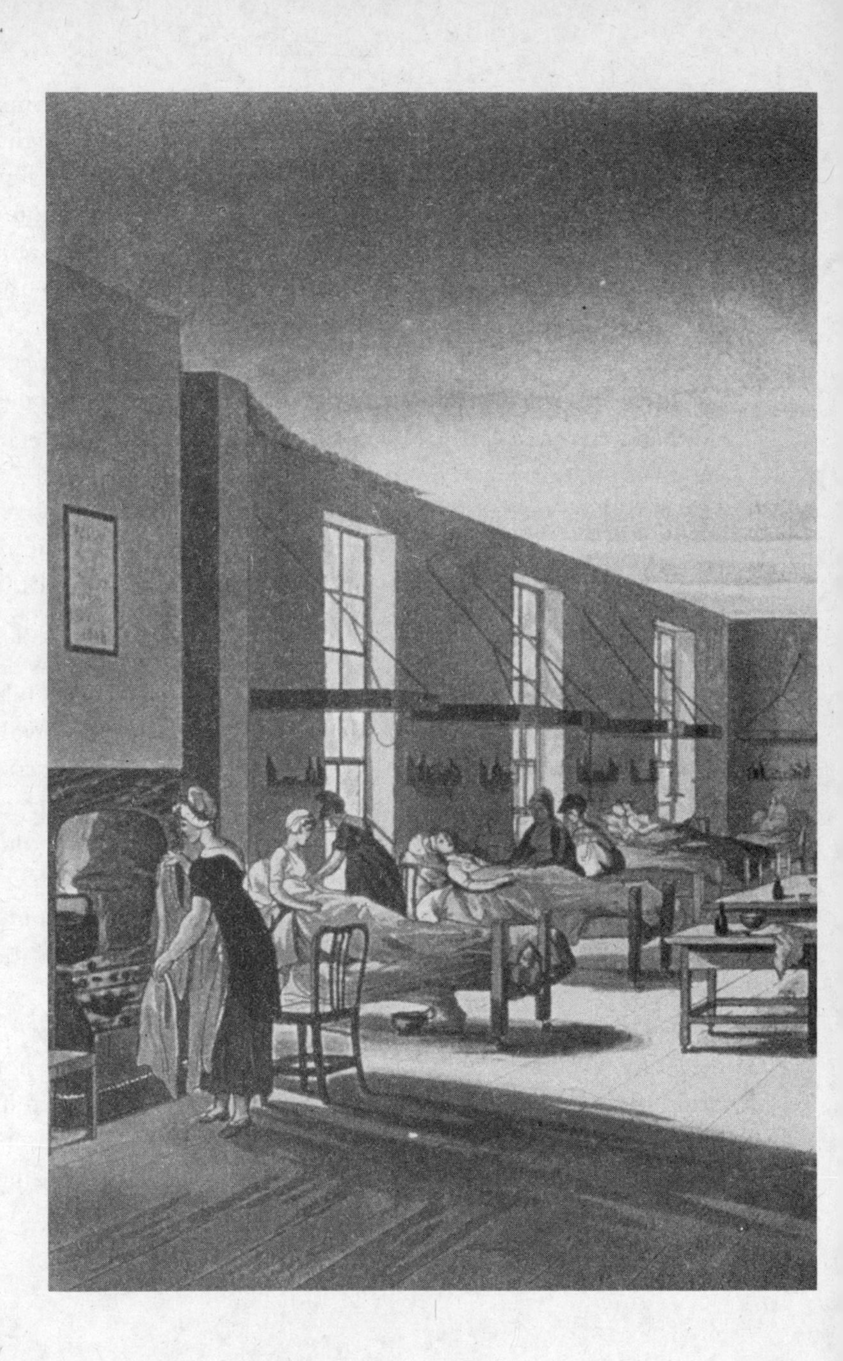

them as gateways to death, because rife with infection, pundits insisted that hospitals did more harm than good. Some held they ought to be burned down every so often and reconstructed, to eliminate the build-up of 'hospitalism' – pyaemia, erysipelas and other contaminations. Fierce debates raged as to how, through better siting, architectural design, ventilation, sanitation and so forth, hospitals could be made safe.

The eighteenth century brought campaigns for hospital reform as part of broad critiques of outmoded, corrupt and harmful institutions. The philanthropist John Howard turned from prison reform to the remodelling of hospitals. He was particularly insistent upon the need for cleanliness and fresh air to combat the deadly miasmic effluvia which he and others blamed for the shocking mortality of gaols and hospitals. In a later era, many, including Florence Nightingale, wanted hospitals removed to the countryside. With such problems, hospitals typically remained for the poor; the wealthy opted to be treated at home. As yet there were no medical procedures exclusive to hospitals: you could be operated upon on the kitchen table, and you gave birth at home.

Medicalization came slowly. Hospitals remained under the control of the lay patrons or religious orders who provided the finance. Nursing, furthermore, had traditionally been provided by religious orders as part of Christian service. In seventeenth-century France the saintly Vincent de Paul set up the Daughters of Charity primarily as a nursing order, and in Catholic Europe, and even North America, nursing remained the vocation of religious orders until recent times. As part of the Revolution's wholesale attack on the Church, religious nursing communities were abolished and charities nationalized in France. By choice and necessity, however, Napoleon largely reverted to the status quo, with hospitals being once more financed by pious donations and staffed by religious orders.

35. 'At the Gates'. The spectres of cholera, yellow fever and smallpox recoil in fear as their way is blocked by a barrier on which is written 'quarantine' and by an angel holding a sword and shield on which is written 'cleanliness'. 1885.

The medical take-over of hospitals came in stages through diverse developments. Increasingly their doors were opened to medical students, and professors with access to clinical beds – like Boerhaave in Leiden – came to use instructive cases as teaching material. In Vienna the hospital reforms carried through in the 1770s by Anton Stoerck led to clinical instruction on the wards, while the success of the Edinburgh medical school owed much to its close links to the city's Infirmary.

With the development around 1800 of new medical approaches based on physical examination, pathological anatomy and statistics (see Chapter 4), the hospital ceased to be predominantly a site of charity, care and convalescence and began to turn into the medical powerhouse it has since become. The new anatomico-clinical medicine, pioneered in Paris by Laënnec at the Hôpital Necker and by Louis at the Hôtel Dieu, was the product of giant public hospitals where direct hands-on experience could be gained, by researchers and students, in abundance. The 'clinic' (as this hospital medicine was to be called) became pivotal to medicine. Hospital facilities were deployed to conduct post-mortems which correlated pathology in the living with internal manifestations after death. Mass observation of patients meant that diseases were identified ontologically as independent entities, rather than being unique to each case, and statistics established representative disease profiles. As well as processing the sick, the nineteenth-century hospital thus became the place *par excellence* where disease could be displayed to students on what became standard ward rounds: being charity cases, the patients could not complain. Further, its morgue was perfect for training students and conducting research.

The nineteenth century also brought a proliferation of specialist hospitals, typically set up by idealistic and ambitious medical men – association with a hospital became a source of professional leverage. By 1860 London alone supported at least sixty-six hospitals and dispensaries catering for a specialism, including the Royal Hospital for Diseases of the Chest (1814), the Brompton Hospital (for tuberculosis: 1841), the Royal Marsden Hospital (for cancer: 1851), the Hospital for Sick Children, Great Ormond Street (1852), and the National Hospital, Queen Square (for nervous diseases: 1860). Similar hospitals were springing up throughout the developed world. Children's hospitals were set up in Paris in 1802, in Berlin (1830), St Petersburg (1834) and Vienna (1837). The Massachusetts Eye and Ear Infirmary was established in 1824, the Boston Lying-In Hospital in 1832, the New York Hospital for Diseases of the Skin in 1836 and scores more.

With the appearance of the modern, medically oriented hospital, nursing too underwent transformation, becoming more professional and acquiring its own career structures and aspirations. In the absence of religious vocations, nursing provision in Protestant countries had always been rather improvised. The stereotypical nurse in England was a slovenly, drunken battle-axe – Dickens's Sairey Gamp and Betsy Prig.

The Deaconess Institute, established in 1836 by Theodore Fliedner, the Lutheran pastor of Kaiserswerth near Düsseldorf, marked a significant advance. It was designed to instruct young ladies to become nurse-deaconesses, a rather superior breed. In 1840 Elizabeth Fry visited it and on her return to London founded the Institute of Nursing.

It was the Crimean War (1853–6), however, which aroused public awareness in England to the need for nursing reform. It produced a heroine in Florence Nightingale. Coming from a comfortable and well-connected background, Miss Nightingale found in nursing the solution to her need to escape her family and use her talents and energies through service. She studied nursing abroad, staying for three months at Kaiserswerth and in

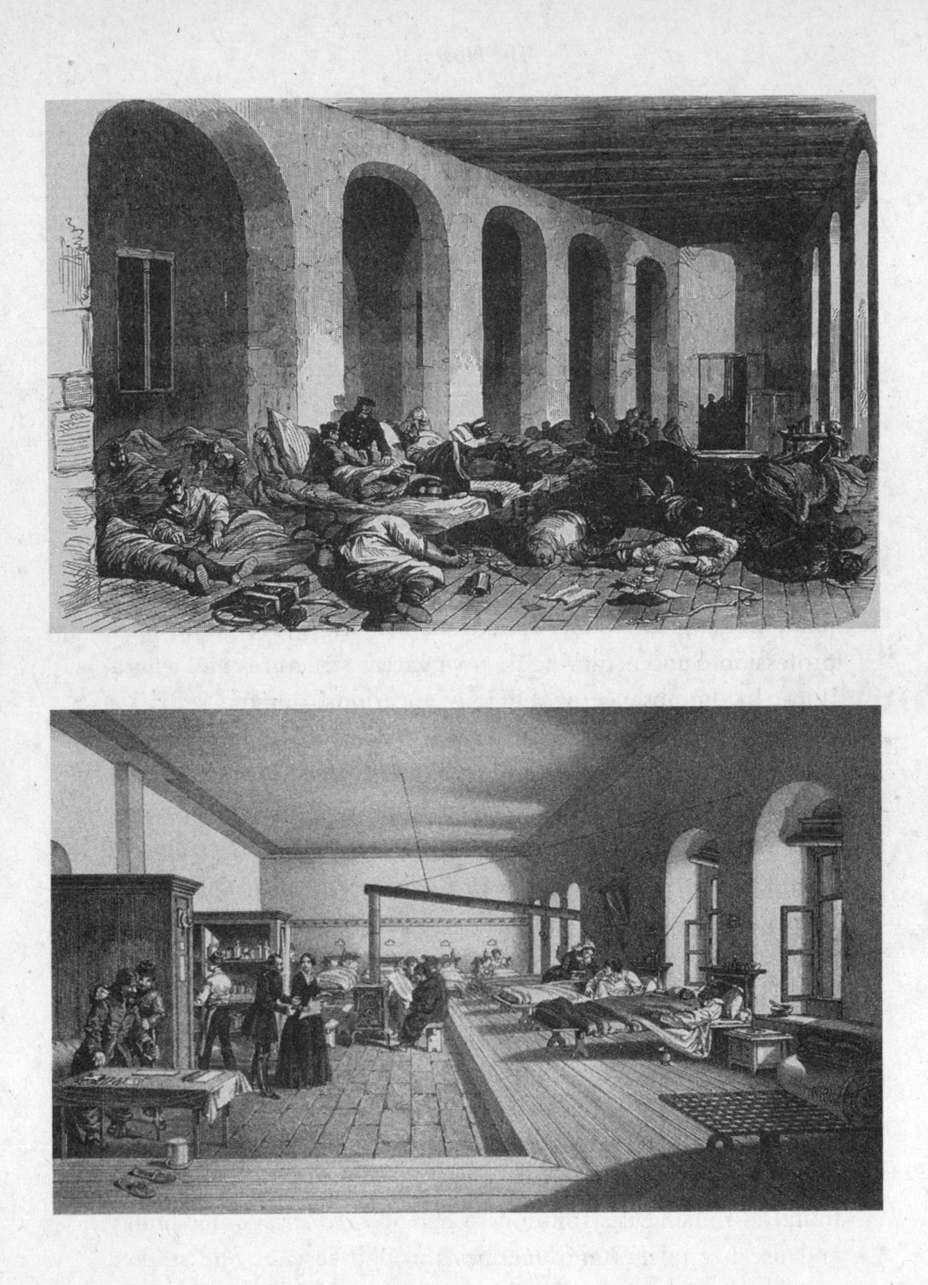

Paris with the Sisters of Mercy. When the shocking dispatches sent from the Crimea by the *Times* journalist W. H. Russell revealed that Britain's wounded soldiers were being looked after by untrained male orderlies, the war secretary Sidney Herbert asked her to put matters right. She arrived with thirty-eight nurses at the barrack hospital at Scutari on the Black Sea. In the teeth of much opposition, she had within six months transformed conditions, and the death rate fell from 40 per cent to 2 per cent.

The extraordinary success of the 'lady with the lamp' produced in 1856 a public subscription to set up a nurses' training scheme. Arrangements were made with St Thomas's Hospital in London, and the first Nightingale nurses started in 1860. Her *Notes on Nursing* and *Notes on Hospitals* stressed hygiene, fresh air, stern discipline, *esprit de corps* and devotion to nursing as a vocation. Nightingale schools became training grounds for superintendents who carried the system throughout Britain and into Australia, Canada, New Zealand and the United States within the next twenty years.

Similar reforms had, indeed, already been moved in the United States by the redoubtable Dorothea Dix who, soon after the outbreak of the Civil War, was appointed Superintendent of the United States Army Nurses. Everywhere, nursing made great strides. Around 1900 Sir William Osler could write that 'the well-trained nurse has become one of mankind's blessings, next to the doctor and the priest – and no less significant than either of them'. His concluding point was true, if it

36. (opposite above) *Appalling conditions of the hospital at Sebastopol during the Crimean War. Wood engraving, 1855.*
37. (opposite below) *Florence Nightingale in much improved conditions at the hospital in Scatari during the Crimean War. Lithograph, E. Walker, 1908.*

somewhat disguised the fact that the (female) nurse was routinely subordinated to the (male) doctor.

From the 1880s the construction of well-equipped and sterile operating theatres where advanced antiseptic surgery was performed helped turn the hospital from a refuge for the indigent into a machine for curing, the saviour of the seriously ill. Alongside free beds for the poor, private wards were built for paying patients.

During the twentieth century surgery steadily became much more intricate, and a host of laboratory tests and other investigations was generated, requiring bulky and costly medical technology (for instance, the massive electrocardiograph) available and usable only in hospital. From the iron lung for polio (1930s) and the dialysis machine (1940s), life-support in various forms became the hospital's business. Ambulance services for accidents and emergencies and blood transfusions further made the hospital the hub of acute care, long before the emergence of designated Intensive Care Units with their ventilators and batteries of monitors. The assumption gained ground that all possibly life-threatening conditions and elaborate medical procedures would self-evidently best be performed within the all-encompassing environment of the hospital. It became normal to be born there in a special maternity unit – and to die there, perhaps in the terminal ward.

A multitude of developments in diagnostics and surgery raised the hospital within the pecking order of medicine and in the public mind. Looking back from 1930 on fifty years of practice in America, Robert Morris reflected:

> One of the very greatest changes that I have observed . . . has been in the attitude of the public toward hospitals. Dread of them was general and well founded before the days of antiseptic surgery. But with its widespread adoption, fear faded rapidly from the lay mind. All over the

world the very name 'hospital' suggested pestilence or insanity; few people would go voluntarily to such a place, no matter how well equipped it was for doing routine work efficiently. To-day, almost everybody with any illness at all serious wishes to go there.

All such changes also caused hospital expenditure to shoot through the roof: by 1950 hospitals were absorbing two-thirds of the resources spent in the USA on health care, and the percentage continued to rise. Especially expensive were technological innovations, from the iron lung and the electron microscope of the 1930s to the million-dollar scanners of the 1970s. Traditionally funded in most countries on an ad hoc and voluntary basis, hospitals found financial problems staring them in the face.

In the United States these were met by the development of business strategies, in conjunction with private insurance schemes which enticed well-off patients to avail themselves fully of hospital facilities. From around the turn of the century, the intimate links forged between top-flight university-based medical education, research centres and philanthropic funding gave a huge boost to the American hospital, both as a bricks and mortar, or glass and concrete, reality and as an icon of medical progress.

The USA enjoyed a hospital boom, and hospitals in turn became the headquarters and power base for the medical élite. By then the profession had gained full control of such institutions and confidence in its leadership was reinforced by an ideology proclaiming that advances in bio-medical science were the pledge of progress. Hospital laboratories would generate medical innovations, which hospital-based medical education would spread through a hierarchy of practitioners and institutions. Patients would reap the benefits. As presented in the movies and on television, up-to-the-minute medicine was medicine that went on in gleaming, streamlined hospitals.

Funds for flagship hospitals, and the research and teaching facilities they housed, were procured from Washington, State governments and philanthropic bodies such as the Rockefeller Foundation. Between the wars, the Rockefeller gave millions of dollars to university departments and hospitals – not just in the USA but in many countries besides – to support the science-based hospital medicine envisaged by the prestigious and influential Flexner Report of 1910 into medical education.

Not surprisingly, with all that support the USA was to lead the world in clinical science, that is hospital-associated research. Awards in the category of 'physiology or medicine' came thick and fast. In 1934 the prize was awarded to George Richards Minot and William Parry Murphy of Boston, together with the pathologist George Whipple, for demonstrating that pernicious anaemia, then a fatal disease, could be successfully treated with a liver diet. In subsequent years Charles Huggins received the prize for his hormone therapy treatment of prostate cancer, Philip Showalter Hench for the introduction of cortisone treatment for arthritis, and Daniel Bovet for the discovery of antihistamines. A glance at the roll of recent Nobel laureates shows that a large proportion of recipients have worked in American hospital medicine, on such problems as cholesterol, retroviruses and transplantation.

The hospital assumed paramountcy in Britain, too, but by a different route – for the links between hospitals, medical education and research remained more oblique. The Second World War led to the effective nationalization of British hospitals – hitherto a motley mixture of public and private, big and small – in anticipation of the massive civilian casualties expected from German bombing. There were two main upshots of this development: cash-strapped hospitals began to count on central government finance, and they became better attuned to cooperation within state-planned schemes. Incorporated from 1948 into the National Health Service, though with élite teaching hospitals left with a measure of self-management, hospitals became its prize, and by far its most costly, sector.

With their skilled coordinated teamwork among many different specialties, in recent times hospitals have been seen as indispensable to modern medical care. Especially in the USA, they converged with huge business corporations. Moving to head the Hospital Corporation of America (Nashville, Tennessee), a former chairman of a fast-food chain explained that 'the growth potential in hospitals is unlimited: it's even better than Kentucky Fried Chicken'. No wonder, because astonishing transformations in scale and through-put were taking place. The annual number of hospital admissions in the United States rose from an estimated 146,500 in 1873 to more than 29,000,000 in the late 1960s. While the nation's population grew five-fold in that period, use of hospitals rose almost two hundred times. In 1909 there were 400,000 beds in the USA; by 1973 there were 1.5 million. In Britain the number of beds per thousand of the population doubled between 1860 and 1940, and then doubled again by 1980. In recent decades the accent has been on stabilizing the number of beds but achieving ever more rapid patient turnover; individual stays are abbreviated in a drive for cost-efficiency.

In our times cutting-edge medicine has been practised in purpose-built hospitals served by armies of paramedics, technicians, ancillary staff, managers, accountants, fund-raisers and other white-collar workers, all held in place by rigid professional hierarchies and codes of conduct. In the light of this massive bureaucratization, it is small wonder that critiques once again emerged. The hospital was no longer primarily denounced, however, as a gateway to death but as a soulless, anonymous, wasteful and inefficient medical factory, per-forming medicine as medicine demanded it, not as the patient needed it.

As a result of policy changes in the field of mental illness, the decades from the 1960s produced massive downscaling and closure of psychiatric hospitals, while hospices were opened to create more sympathetic environments for the terminally ill. Whether, for its part, the general medicine of the future needs, or can afford, the ever-expanding hospital complex remains unclear. Today's huge general hospitals may soon seem medicine's dinosaurs. Will they go the way of the lunatic asylums?

CHAPTER EIGHT

Medicine in Modern Society

> There is virtually no limit to the amount of health care an individual is capable of absorbing.
>
> Enoch Powell, British Minister of Health

Through most of its history Western medicine was a small-scale affair, based on the face-to-face encounter between a sick person and a healer, amateur or professional, regular or quack. Practitioners were mainly self-employed, and the patient–doctor relation involved a voluntary, private and confidential transaction. Other healing set-ups, such as outpatient charities and religious shrines, made much of the personal touch.

All this has changed. Modern health care has turned into a colossal service industry, in both state and private sectors; in many nations it claims a greater share of the gross national product than any other item – these days, a staggering 15 per cent in the USA. Critics call it a juggernaut, an institution out of control, or at least one driven less by patient need than by profit and professional power. The personal touch, so essential to healing, has been lost, claim the millions who in turn have lost faith in Western scientific medicine.

This transition from small man to corporate enterprise is the result, in part, of those giant strides in basic and clinical research, and the pharmacological and surgical revolutions discussed in the preceding chapters. Back in the 1850s Claude Bernard was able to fund his research out of his wife's dowry – the marriage was not a happy one! – while the George Sumner

38. *Doctor and surgeon discussing a patient.* Punch *cartoon, 1925.*

Huntington (1851–1916) who elucidated Huntington's chorea was an obscure American country practitioner; all the tools of his trade fitted into his saddle-bag. But even his contemporary, the bacteriologist Robert Koch, who also started out as a small country doctor, ended up lording it over several palatial research institutions; and since then the iron law has been expansion, capital investment, bureaucratization, commodification and the economics of scale and the division of labour. Regular medicine is nowadays unthinkable without its research centres and high-tech teaching hospitals. The medical machine has acquired an extraordinary momentum.

And if, in the increasingly intricate medical division of labour, physicians, who once ruled the roost, remain high in status, they have today become but a cog in the machine. Of the 4.5 million employees involved in health care in the USA in the late twentieth century (5 per cent of the entire labour force), only about one in seventeen was a practising physician. Perhaps nine out of ten of those employed in the modern medical enterprise never directly treat the sick. Two hundred years ago, by contrast, there were practically no medical administrators or other 'non-combatants'.

This systematization of modern health-care delivery has gone hand-in-hand with transformation in its remit and objectives. Formerly the physician simply treated the sick man, woman or child as best he could; over time, however, medicine asserted, and was called upon to play, a much greater and more proactive role in the welfare of the citizen and in the healthy ordering of society. Within emergent welfare or therapeutic states, medicine staked its claims for a mission within the home, the office and the factory, in law courts and schools, in the city and in the military. The more scientific and effective medicine grew, the more the public itself, its political representatives, and the media, looked to its benevolent potential, casting the healing arts as a fairy godmother who would, it was hoped, grant everybody's wishes.

In advanced market societies medicine is, moreover, a commodity for which – thanks to the growth of surplus income – demand has constantly been rising. And ever since the wily Chancellor Otto von Bismarck set up state-run medical insurance in newly unified Germany in 1883, politicians for their part have been able to look to improved health care as a carrot to dangle before the electorate. Votes were to be had not just in bread and circuses but in beds and surgery.

Whilst growing, before the twentieth century the state's role in all such matters remained rather ad hoc. Statutory medical provision tended to be limited to individual problems (for instance, the policing of communicable disease). By 1900 medical professionals were everywhere licensed by law, but nowhere did the state actually ban irregulars, and medical ethics ensuring good medical practice were largely left to professional self-regulation. In industrialized nations public-health legislation had entered the statute book to control matters like sewage, sanitation and smallpox. Yet, above all in the USA, health care remained a patchwork of voluntary, religious and charitable initiatives, as was relief for the old and needy, while medicine for those who could afford to pay was still as much a private transaction as buying furniture or hiring a music tutor.

All this was to change, continuously if unevenly, in the twentieth century. It became accepted that the efficient functioning of exceedingly intricate industrial economies, in peace and war, required a population which was no less healthy than literate and law-abiding; and in democracies where workers had become voters, provision of health services, as Bismarck had seen, became one means of pre-empting discontent and revolution.

Improved health also weighed heavily in twentieth-century propaganda wars. Fascist Italy, Nazi Germany and the Communist USSR all worshipped at the altars of health and fitness. While diagnosing the social pathogens said to sabotage national well-being and seeking to eradicate the 'cancer' of the Jews, the Nazis exalted macho workers, fertile mothers and bonny children, encouraged the cult of physical fitness through hiking, paramilitary drill, sport and sun-bathing, and launched the first anti-smoking campaigns. In any case, be they democratic or totalitarian, the hands of great powers were tied when it came to health matters: world wars required massive injections of public money and resources into centralized health services to keep soldiers in the field and sustain civilian morale alike.

As the twentieth-century ship of state (no matter what flag it flew under) took health on board, medical thinkers were spelling out a new mission for the profession. Conventional one-to-one clinical medicine, argued reformers, was itself handicapped and short-sighted. Why wait till people fell sick? Was not prevention better than patching? Surely it was far better to discover what produced disease in society in the first place and then, directed by statistics, sociology and the exciting new discipline of epidemiology, take measures to build positive health. In a rational, democratic and progressive society medicine should have a voice, it should lead not follow. It should get to the root of the pathological tendencies in the community and eradicate them through far-sighted policies, laws, education and specific agencies and practices: screening, testing, health information, ante-natal care and infant welfare.

Because it dealt with ill-health only at the individual, clinical level, conventional medicine was criticized as myopic – it was like endlessly mopping up a flooded bathroom floor rather than fixing a leaking tap. Health needed to be grasped as an expression of the vitality of the collectivity. Moreover, it could not be combated ad hoc, but only through planned interventions. Such views, sometimes called 'social medicine', were widely embraced in Europe, and to a certain degree in North America, among planners and civil servants, Socialists and Marxists, progressive doctors, medical rationalists, and not least patriots preoccupied with ensuring national mastery in a cut-throat Darwinian political cockpit whose very law was biomedical – prosper or perish.

This call for medicine to modernize was affected by recognition that the disease environment itself was changing. The prevalent diseases, epidemiologists were discovering, were ceasing to be the classic air-, water- and bug-borne infections which had bedevilled the early industrial age: cholera, typhus, typhoid and so forth. In place of the old disease empire, chronic disorders now assumed a heightened prominence. Medicine had to turn its attention to a morass of deep-seated and pervasive dysfunctions hitherto rather neglected: delicate infants, backward children, anaemic mothers, clerks with ulcers, sufferers from arthritis, back pain, strokes, inherited ailments, depression and other neuroses, and all the ills of old age which greater longevity was bringing to the fore.

To counter all this suffering, hardship and waste, medicine, it was asserted, must become a positive and systematic enterprise, undertaking planned surveillance of seemingly healthy and normal people as well as the sick, tracing cohorts from infancy to old age, charting the incidence of inherited, chronic and constitutional conditions, and plotting ill-health against such variables as income, education, class, diet and housing. Disease thus became conceptualized in the twentieth century as a social no less than a biological phenomenon, to be understood statistically, sociologically, psychologically – and politically.

The twentieth century generated a welter of programmes and policies devoted to rectifying this newly uncovered social pathology and to improving the people's health. Their underlying ideologies ranged from the Socialist Left (state medicine should implement social justice and help the underprivileged) to the

Fascist Right (nations must defend themselves and their colonies against social 'germs'). Either way, the liberal-individualist Hippocratic model of medicine as a sacred private contract between patient and bedside doctor seemed to many as *passé* as was Smithian *laissez-faire* political economy in the age of the Slump and John Maynard Keynes.

New philosophies of health thus embraced positive and hopeful visions of the socialization of medicine and the medicalization of society. Buoyed up by the successes of bacteriology, tropical medicine and the surgical revolution, confidence soared about what medicine and health care might achieve. In a world torn by war, violence, class struggle and economic depression, would not medicine at least be a noble force for good, not least in the underdeveloped world? The benefits were obvious; the limitations would surface only later.

Over the centuries various push-and-pull mechanisms had in actuality been drawing medicine, albeit slowly and in a piecemeal fashion, into the public domain, be it the state or the market. Doctors were always liable to be called upon in time of emergency, particularly plague and war. And in the nineteenth century new growth-points arose in public medicine, notably the need to cope with the looming threat of the sick poor and the environmental hazards caused by industrialization.

Driven by humanitarianism and prudence, medical measures were devised to alleviate the afflictions of the masses. The nineteenth century brought dispensaries and hospitals for the sick poor supported by charity (religious or secular) and public subsidy. After the new Poor Law of 1834, England's workhouse infirmaries provided beds for hordes of paupers.

Confronting the rampant diseases of industrial society, the sanitary movement preached clean water and good drains, physical and moral cleanliness; and in some nations, notably Britain, it won coercive judicial powers. States and cities created appointments for doctors as medical officers of health, public analysts, factory inspectors, forensic experts, prison doctors and asylum superintendents. Doctors might find employment within the public sector, while fearful of compromising their independence.

Meanwhile, the market provided beckoning opportunities for some. From the progressive era onwards, American medicine particularly proved inventive and energetic in promoting new specialties and commercial arrangements, providing wider services and diagnostic tests, and tapping new sources of custom and income. Medicine became another line of business, and business was booming.

Especially in big cities, American private practitioners discovered the advantages of behaving like their fellow lawyers or businessmen, setting up downtown offices and providing impressive facilities – developments almost unknown in the UK. Progressive in the use of telephonists and technicians, X-ray machines and chemical laboratories, physicians attracted patients by radiating confidence. By 1929 the Mayo Clinic in Rochester, Minnesota had become a huge outfit with a staggering 386 physicians on its books and 895 lab technicians, nurses and other workers. Housed in a fifteen-storey building, the clinic had 288 examining rooms and twenty-one laboratories.

In Britain the garden seemed less rosy for ordinary private doctors; and with politicians mooting schemes for health assurance and state medicine, it was easy for them to feel left out in the cold. A quandary thus faced the profession in 1911 when the Liberal politician Lloyd George launched his National Insurance plan, modelled along Bismarckian lines. This scheme proposed providing state medical insurance for the working classes, with contributions shared between individual, employer and the state. Insured workers would receive approved medical treatments from a 'panel doctor' and a cash benefit for the first thirteen weeks of sickness (men would get more than women). There were certain restrictions – hospital costs were not met, except in the case of tuberculosis sanatoria – and the families of insured parties were excluded, though there was a maternity grant (babies were the nation's future). It was a measure devised to be popular with the electorate while doing something to ameliorate the wretched health of workers, as exposed by the recruitment crisis in the recent Boer War.

Initially practitioners were up in arms: they would not be reduced to the state's dogsbodies! In the event, however, most opted to become panel doctors and found that their new relationship with the state was secure and remunerative. National Insurance widened the gap in Britain between the general practitioner and the hospital consultant, and this was to have long-term repercussions for the structure of the profession. But it also consolidated a valued relationship between the sick and their GPs, who were appreciated because they were a reassuringly tangible presence.

In the interwar years the typical family in advanced societies was the focus of innovative public medicine and health policies, but the precise nature of the arrangements varied from state to state. After Prime Minister Lloyd George's ringing promise during the First World War to create a land fit for heroes, the victorious British experienced post-war poverty, unemployment and sickness. In line with his belief that 'at no distant date, the state will acknowledge a full responsibility in the matter of provision for sickness', a Ministry of Health was established in 1919 – but that proved a substitute, rather than a springboard, for further action.

After the 1917 Bolshevik Revolution the USSR moved to a salaried state medical and hospital service which prized science and expertise. Free and universal treatment financed out of taxes was a right. If its standard was not uniformly high, this nevertheless represented an enormous leap forward. Germany for its part continued to operate its established Bismarckian state-regulated insurance scheme for workers, administered, like its British equivalent, through voluntary friendly societies or employer schemes. Excluded from state benefits, some of the middle classes pre-paid for doctors through private or occupational insurance schemes. In France a state insurance system reimbursed patients for the fees of physicians, and gave them free choice of doctor and hospital. Public hospitals, however, remained cash-starved and low-grade, and the insured flocked to private ones. While steps were taken to benefit mothers and babies, in line with pronatalist attempts to boost population, the ethos of economic liberalism remained strong in France, protecting the freedom of both patients and doctors, and shying away from German compulsory state medical insurance. Not until 1930 was a social insurance law finally enacted.

In the USA health insurance became a lasting political football. Initially the American Medical Association kept its options open, but in the chauvinistic atmosphere after the First World War, when everything German and Soviet became vilified, attitudes hardened and the AMA came out against. 'Compulsory Health Insurance,' declared one Brooklyn physician, 'is an Un-American, Unsafe, Uneconomic, Unscientific, Unfair and Unscrupulous type of Legislation supported by . . . Misguided Clergymen and Hysterical Women.' The *Journal* of the AMA feared such insurance would reduce Americans to automatons. Turning more conservative, the AMA resisted the Sheppard-Towner Act, which provided federal subsidies for maternal and child health programmes, and opposed the establishing of Veterans Hospitals in 1924 – both would take the bread out of the private physician's mouth.

Designed to spend the nation out of the Depression, President

Franklin Roosevelt's New Deal seemed to be leading America along the road towards a national health programme, and many New Deal agencies were indeed involved in health. The severity of the Depression and the popularity of FDR – himself a polio victim – forced the AMA to temper its views.

During the Depression, when many could no longer afford to pay medical fees and the once-buoyant hospital sector plunged into crisis, hospitals began to introduce voluntary insurance schemes to cushion their users, and commercial companies moved into the insurance market. This led to the Blue Cross (hospital) and Blue Shield (medical and surgical) pre-paid programmes. If initially suspicious, the AMA soon tempered its opposition – such voluntary schemes suited its outlook better than compulsory federal ones.

The result was that health insurance became big business. In the twenty years after 1940, private insurance enjoyed explosive growth, and the insurance model captured American private medicine. Middle-class families, or often their employers, paid for primary and hospital care through insurance schemes, and physicians and hospitals competed for buoyant custom.

Medical politics, meanwhile, had taken an altogether different turn in Germany. Founded in 1908, the *Archiv für Rassenhygiene* (Archive of Race-hygiene), the main organ of the German eugenics movement, demanded action to halt the 'biological and psychological deterioration' of the Aryan race. In his *Mein Kampf* (1927), Adolf Hitler, Chancellor from 1933, demonized Jews, gypsies and other groups as enemies of the master race, and Nazi medicine in due course defined some of those non-Aryans as subhuman. The anti-Semitism which culminated in the Holocaust received the blessing of prominent physicians and psychiatrists, organized through the Nazi Physicians' League.

Doctors and scientists eagerly promoted and participated in

such Nazi policies as the sterilization of the 'genetically unfit'. Physicians sterilized nearly 400,000 mentally handicapped, epileptics and alcoholics even *before* the outbreak of war in September 1939. Thereafter, 'mercy deaths' became routine at mental hospitals: between January 1940 and September 1942, 70,723 mental patients were gassed. Some were victims of Nazi programmes of human experimentation. The 'final solution' of the 'Jewish problem' was given full medical rationalization.

Doctors also pursued human experimentation in Japan. In 1936 a medico-scientific centre, led by Dr Shiro Ishii, was set up in Pingfan in northern Manchuria, then under Japanese military occupation, to pioneer bacterial research. It produced enough lethal microbes – anthrax, dysentery, typhoid, cholera and bubonic plague – to wipe out the human race several times over; some of these were tested on the local population.

A post-war reaction against such perversions was the international ethical movement for medicine, one of whose fruits was the Nuremberg Code (1947). Though it failed to define genocide as a crime, the Code was nevertheless meant to ensure medical research could never again be abused. Its principles were further refined in the Declaration of Helsinki on medical research (1964), which differentiated between therapeutic experiments (clinical research combined with professional care) and non-therapeutic experiments (ones of no benefit to the subject).

In Britain the idealism and optimism which greeted the end of the Second World War brought a unique reorganization of medical services. Its blueprint was the *Beveridge Report on Social Insurance and Allied Services* (1942) which declared war on the 'five giants' threatening society: Want, Ignorance, Disease, Squalor and Idleness. It proposed that a new health service should be available to everyone according to need, free at the point of service, without payment of insurance contributions and irrespective of economic status. All means tests were to be abolished. It was a noble vision.

Winning a landslide victory at the general election of 1945, the Labour Party set about implementing the Report; the appointed day for its inauguration was 5 July 1948. Among the key changes, Aneurin Bevan, the Minister of Health, nationalized all hospitals, municipal as well as charity. No friend of local government, he wanted hospitals – now recognized, as in America, as flagship institutions – to be under central control. The reorganization, in which the government became responsible for 1,143 voluntary hospitals with over 90,000 beds, together with 1,545 municipal hospitals with 390,000 beds, amounted to the most far-reaching action relating to hospitals ever brought about in a Western nation. Overall, however, the NHS did not revolutionize medicine – indeed it perpetuated the old division between consultants and GPs, still then mainly operating in single-handed practices: thereafter the consultants had the hospitals but the GPs retained the patients.

An acceptable standard of medicine was now for the first time readily available to all. The NHS system proved efficient, fairly equitable, and for a long time enormously popular. Hopes that better treatment would lead to a need for less medicine and hence to reduced expenditure proved, however, fanciful. Likewise, bitter experience showed that socialized medicine did not, in the event, reduce the marked inequalities of health between the affluent and the poor. By the close of the century, the long-term penny-pinching in capital investment – always probable with a centrally funded system designed to keep taxes low – was jeopardizing the future of a remarkably successful experiment and undermining public confidence.

Broadly comparable developments to the NHS had occurred, or were to follow, in other British-influenced nations, such as New Zealand. Canada, too, took the path of socialized medicine, at a later date. Saskatchewan began its Medical Care Insurance and Hospital Services Plan in 1962, enabling residents to obtain insurance covering many medical services. This government-administered programme was funded by an annual tax and federal funds. In 1967 a central Medical Care Act coordinated the system across the nation.

As western Europe recovered from the devastation of the war and moved during the 1950s into a new age of affluence, a diversity of state-supported medical schemes took shape. Sweden established its system of medical care and sickness benefit insurance in 1955. West Germany continued to use sick-funds which reimbursed doctors, while France still relied on state welfare benefits through which patients were refunded for most of their medical outlays.

Meanwhile the USA continued its own way. As already noted, from the 1930s those who could took out private health insurance, helped by tax-deductible occupational schemes. With the fee-for-service system entrenched, physicians and hospitals competed to offer superior services – more check-ups, better tests, the latest procedures, a menu of elective surgery and so forth. Costs inevitably soared, as did profits, and when President Truman mooted a national health programme in 1948, the AMA campaigned effectively against it.

Despite the nation's ideological commitment to private medicine, the American government in reality shouldered a growing proportion of health care. Federal government provided direct medical care to millions through the Armed Services, the Veterans Administration, the Public Health Service and the Indian Health Service.

Complementing established private health insurance schemes such as Blue Cross came the Health Maintenance Organizations (HMOs), originating with the Kaiser Foundation Health Plan in California. By 1960 that scheme was providing comprehensive medical care to over half a million subscribers, while by 1990 it employed 2,500 physicians in fifty-eight clinics and twenty-three hospitals. Subscribers to HMOs – cheaper than regular insurance – paid monthly dues entitling them to comprehensive medical care.

The disparity between lavish provision for well-off insured families and the plight of the poor and the old became more glaring. This injustice proved a source of national embarrassment and a campaigning platform for the Democratic Party. Capitalizing on the wave of idealist sentiment following President Kennedy's assassination, in 1965 his successor Lyndon B. Johnson made medical care a social security benefit through Medicaid, set up alongside Medicare, a parallel health-and-care plan for the old. Both schemes proved inflationary because providers were reimbursed on the standard fee-for-service basis.

Health became a key growth sector in the American economy, encompassing the pharmaceutical industry, manufacturers of diagnostic apparatus, laboratory instruments and therapeutic devices, in addition to medical personnel, hospital teams and their penumbra of corporate finance, insurers, lawyers, public relations firms and accountants. Expenditure has continued to rise, quite disproportionately to measurable improvements in health.

The apparent paradox that the world's most prosperous nation (and overall one of the healthiest) was inexorably spending ever more on medicine drew criticism from various quarters. Conservatives denounced Medicare and Medicaid as a blank cheque, corrupting to consumers and providers alike within a flawed high-cost medi-business system geared principally to benefiting the supply side. Consumers challenged professional and commercial monopolies, setting up patients' groups and stressing patients' rights. The medical establishment itself came under attack from the time of the 1960s populist counter-culture backlash against scientific and technological arrogance. Disasters with new drugs, notably thalidomide, were seen as proofs of technical failure and professional malpractice. Other protests grew louder. Critics of vast, impersonal mental hospitals campaigned for their closure. Feminists for their part lambasted patriarchal medicine, as evidenced in the hospitalization of normal births, and through such slogans as 'Our Bodies, Our Selves' reasserted control over their own bodies. And exposés of out-of-control health costs further spotlighted the predicament of those excluded from its benefits. By 2000 some 40 million Americans had no medical insurance – almost one in every six citizens under the age of sixty-five.

Criticism of the medical system grew fiercer throughout the West in the last decades of the twentieth century. Was health care cost-effective? Was it equitable? Was it safe? How could the public be protected against medical malpractice? There was an irony in this as people at large were leading longer, healthier lives than ever before. This erosion of confidence led many to try alternative forms of healing which seemed more patient-friendly. But neither in North America nor in those European nations suffering from the crises of the welfare state has this tide of criticism produced structural reforms, merely a ragbag of cost-capping initiatives, accounting and managerial strategies and short-term economies. The reform of the American health care system, promised at the beginning of the Clinton administration in 1992, came to absolutely nothing. And meanwhile the appropriateness of much of the medicine exported by the West to the Third World has come increasingly into question. Despite the global eradication of smallpox, in many underdeveloped nations malaria, tuberculosis and AIDS rampage out of control.

During the twentieth century health care became integral to the machinery of industrialized society. The consequences are not easy to evaluate. The enormous inequalities of health between rich and poor, revealed by nineteenth-century statisticians, remain, while the disparities between the health standards of the First and Third worlds have blatantly increased. Modern medicine at its best possesses unique capacities to keep individuals alive, healthy and free of pain. Its contribution to the broader health of humankind remains more questionable. Many believe that investment in public health, environmental hygiene and better nutrition would do far more for the health of Third World nations than sophisticated clinical medicine programmes.

Meanwhile environmental improvements and better living standards today contribute more than curative medicine in guaranteeing the longer lifespans which are now taken for granted. And medicine is making only slow inroads against the diseases of ageing. In the light of these factors, the role and scope of medicine in advanced states seem destined to change in the twenty-first century as the accent shifts from overcoming disease to the fulfilment of life-style wishes, bodily enhancement and further extensions of life. Thus poised, medicine may be on the brink of one of the greatest transformations in its long and chequered history. But right now, after the golden age of some generations back, the public climate is one not of optimism but of new-millennial anxiety.

Further Reading

Here are some suggestions for readers wishing to delve further into the topics covered in this book.

Fine lengthier general histories of medicine include:

Lawrence Conrad, Michael Neve, Vivian Nutton, Roy Porter and Andrew Wear, *The Western Medical Tradition: 800BC to AD1800* (Cambridge: Cambridge University Press, 1995) [takes the story up to 1800]

Jacalyn Duffin, *History of Medicine: A Scandalously Short History* (Toronto: University of Toronto Press, 1999) [this is actually 430 pages long!]

Nancy Duin and Jenny Sutcliffe, *A History of Medicine: From Prehistory to the Year 2020* (London and New York: Simon and Schuster, 1992) [Well written and illustrated]

Mirko D. Grmek (ed.), *Western Medical Thought from Antiquity to the Middle Ages* (Cambridge, MA: Harvard University Press, 1998)

Thomas S. Hall, *History of General Physiology 600 B.C. to A.D. 1900*, 2 vols (Chicago: University of Chicago Press, 1975)

Robert P. Hudson, *Disease and its Control: The Shaping of Modern Thought* (Westport: Greenwood Press, 1983)

Irvine Loudon (ed.), *Western Medicine: An Illustrated History* (Oxford: Oxford University Press, 1997)

Lois N. Magner, *A History of Medicine* (New York: Marcel Dekker, 1992)

—, *A History of the Life Sciences*, 2nd ed. (New York: M. Dekker, 1994)

Roy Porter (ed.), *The Cambridge Illustrated History of Medicine* (Cambridge: Cambridge University Press, 1996)

I have drawn in this book upon materials discussed at greater length in my *'The Greatest Benefit to Mankind': A Medical History of Humanity* (London: HarperCollins, 1997), which may be consulted for further details and bibliography. A recent attempt to set out my thinking on the social meaning of sickness and medicine is *Bodies Politic: Disease, Death and the Doctors in Britain: 1650–1914* (London: Reaktion Books, 2001).

WORKS OF REFERENCE

Jessica and Elmer Bendiner, *Biographical Dictionary of Medicine* (New York: Facts on File, 1990)

Colin Blakemore and Sheila Jennett (eds.), *The Oxford Companion to the Body* (Oxford: Oxford University Press, 2001)

W. F. Bynum and Roy Porter (eds.), *Companion Encyclopedia of the History of Medicine*, 2 vols (London: Routledge, 1993)

Stephen Lock, John Last and George Dunea (eds.), *The Oxford Illustrated Companion to Medicine* (Oxford: Oxford University Press, 2001)

Roderick E. McGrew, *Encyclopedia of Medical History* (New York: McGraw-Hill, 1985)

Leslie T. Morton, *A Medical Bibliography (Garrison and Morton): An Annotated Check-list of Texts Illustrating the History of Medicine*, 4th ed. (Aldershot, Hants: Gower, 1983)

Leslie T. Morton and Robert J. Moore, *A Bibliography of Medical and Biomedical Biography* (Aldershot: Scolar Press, 1989)

Enjoyable anthologies include:

D. J. Enright (ed.), *The Faber Book of Fevers and Frets* (London: Faber, 1989)

Richard Gordon, *The Literary Companion to Medicine: An Anthology of Prose and Poetry* (London: Sinclair-Stevenson, 1993)

I have deliberately avoided in this book involving myself in historiographical controversies as to the best approaches to the history of medicine. For a lively and up-to-date introduction, see Ludmilla Jordanova, 'The Social Construction of Medical Knowledge', *Social History of Medicine*, viii (1995), 361–82.

PREFACE

For the subjective side of disease, death and medicine, see:

Philippe Ariès, *The Hour of Our Death* (London: Allen Lane, 1981)
Sander L. Gilman, *Health and Illness: Images of Difference* (London: Reaktion Books, 1995)
C. Helman, *Culture, Health and Illness: An Introduction for Health Professionals* (Bristol: Wright, 1984) [insights by a medical anthropologist]
David B. Morris, *Illness and Culture in the Postmodern Age* (Berkeley: University of California Press, 1998)
Roselyne Rey, *History of Pain*, tr. by Elliott Wallace and J. A. and S. W. Cadden (Paris: Éditions la Découverte, 1993)
Susan Sontag, *Illness as Metaphor* (New York: Farrar, Straus & Giroux, 1978; London: Allen Lane, 1979)
—, *AIDS and its Metaphors* (Harmondsworth: Allen Lane, 1989)

For non-Western attitudes, see:

John Hinnells and Roy Porter (eds.), *Religion, Health and Suffering* (London: Kegan Paul, 1999)
A. Kleinman, *Patients and Healers in the Context of Culture: An Exploration of the Borderland between Anthropology, Medicine, and Psychiatry* (Berkeley: University of California Press, 1980)

CHAPTER ONE: Disease

Encyclopedic is:

Kenneth F. Kiple (ed.), *The Cambridge World History of Human Disease* (Cambridge: Cambridge University Press, 1993)

Fine general surveys are offered in:

Alfred W. Crosby, *Ecological Imperialism: The Biological Expansion of Europe, 900–1900* (New York: Cambridge University Press, 1986)

Laurie Garrett, *The Coming Plague: Newly Emerging Diseases in a World Out of Balance* (Harmondsworth: Penguin, 1994)

Arno Karlen, *Man and Microbes* (New York: Putnam, 1996)

W. H. McNeill, *Plagues and Peoples* (Oxford: Anchor Press, 1976)

For more specific diseases, see:

Alfred W. Crosby, *The Columbian Exchange, Biological and Cultural Consequences of 1492* (Westport, CT: Greenwood Press, 1972)

Thomas Dormandy, *The White Death: A History of Tuberculosis* (London: Hambledon Press, 1999)

M. Durey, *The Return of the Plague: British Society and the Cholera 1831–2* (Dublin: Gill & Macmillan, 1979)

Richard J. Evans, *Death in Hamburg: Society and Politics in the Cholera Years 1830–1910* (Oxford/New York: Oxford University Press, 1987)

Robert S. Gottfried, *The Black Death: Natural and Human Disaster in Medieval Europe* (New York: The Free Press, 1983)

Mirko D. Grmek, *History of AIDS: Emergence and Origin of a Modern Pandemic*, tr. by Russell C. Maultiz and Jacalyn Duffin (Princeton: Princeton University Press, 1994)

D. Hopkins, *Princes and Peasants: Smallpox in History* (Chicago: University of Chicago Press, 1983)

Gina Kolata, *Flu: The Story of the Great Influenza Pandemic of 1918 and the Search for the Virus that Caused it* (New York: Farrar, Straus & Giroux, 1999)

Randolph M. Nesse and George C. Williams, *Evolution and Healing. The New Science of Darwinian Medicine* (London: Weidenfeld and Nicolson, 1995)

CHAPTER TWO: Doctors

For general accounts, see:

Sherwin Nuland, *Doctors: The Biography of Medicine* (New York: Knopf, 1988)

Edward Shorter, *Doctors and their Patients: A Social History* (New Brunswick: Transaction, 1991)

John Cule, *A Doctor for the People: 2000 Years of General Practice in Britain* (London: Update, 1980)

More specific accounts in chronological order of subject:

J. Worth Estes, *The Medical Skills of Ancient Egypt* (Canton, MA: Science History Publications, 1989)

Carole Reeves, *Egyptian Medicine* (Princes Risborough, Bucks: Shire Publications, 1991)

James N. Longrigg, *Greek Rational Medicine* (London: Routledge, 1993)

E. D. Phillips, *Greek Medicine* (London: Thames and Hudson, 1973)

V. Nutton, 'What's in an Oath?', *Journal of the Royal College of Physicians of London*, xxix (1995), 518–24

Helen King, *Hippocrates' Woman: Reading the Female Body in Ancient Greece* (London: Routledge, 1998)

Ralph Jackson, *Doctors and Diseases in the Roman Empire* (Norman, OK: University of Oklahoma Press, 1988)

David Gentilcore, *Healers and Healing in Early Modern Italy* (Manchester: University of Manchester Press, 1998)

Mary Lindemann, *Medicine and Society in Early Modern Europe* (Cambridge: Cambridge University Press, 1999)

Andrew Wear, *Knowledge and Practice in English Medicine 1550–1680* (Cambridge: Cambridge University Press, 2000)

Lucinda McCray Beier, *Sufferers and Healers: The Experience of*

Illness in Seventeenth-Century England (London: Routledge & Kegan Paul, 1987)
Laurence Brockliss and Colin Jones, *The Medical World of Early Modern France* (Oxford: Clarendon Press, 1997)
Christopher Lawrence, *Medicine in the Making of Modern Britain, 1700–1920* (London and New York: Routledge, 1994)
Roy and Dorothy Porter, *In Sickness and in Health: The British Experience 1650–1850* (London: Fourth Estate, 1988)
Dorothy and Roy Porter, *Patient's Progress: Doctors and Doctoring in Eighteenth-Century England* (Cambridge: Polity Press, 1989)
Irvine Loudon, *Medical Care and the General Practitioner 1750–1850* (Oxford: Clarendon Press, 1986)
Thomas Neville Bonner, *Becoming a Physician: Medical Education in Britain, France, Germany and the United States, 1750–1945* (New York and Oxford: Oxford University Press, 1995)
John Harley Warner, *The Therapeutic Perspective: Medical Practice, Knowledge and Identity in America, 1820–1885* (Cambridge, MA: Harvard University Press, 1986)
Anne Digby, *Making a Medical Living: Doctors and their Patients in the English Market for Medicine, 1720–1911* (Cambridge: Cambridge University Press, 1994)
Anne Digby, *The Evolution of British General Practice 1850–1948* (Oxford: Oxford University Press, 1999)

Quack and fringe medicine are covered in:

Norman Gevitz, *Other Healers: Unorthodox Medicine in America* (Baltimore and London: Johns Hopkins University Press, 1988)
Phillip A. Nicholls, *Homoeopathy and the Medical Profession* (London and New York: Croom Helm, 1988)
Roy Porter, *Quacks: Fakers and Charlatans in English Medicine* (Stroud: Tempus, 2000)
Mike Saks (ed.), *Alternative Medicine in Britain* (Oxford: Clarendon Press, 1991)
James Harvey Young, *The Medical Messiahs: A Social History of Health Quackery in Twentieth-Century America* (Princeton, NJ: Princeton University Press, 1967)

For women and medicine, see:

Thomas Neville Bonner, *To the Ends of the Earth: Women's Search for Education in Medicine* (Cambridge, MA/London: Harvard University Press, 1992)

Rosemary Pringle, *Sex and Medicine: Gender, Power and Authority in the Medical Profession* (Cambridge: Cambridge University Press, 1998)

CHAPTER THREE: The Body

For broad cultural attitudes, see:

M. Feher (ed.), *Fragments for a History of the Human Body*, 3 vols (New York: Zone, 1989)

Martin Kemp and Marina Wallace, *Spectacular Bodies: The Art and Science of the Human Body, from Leonardo to Now* (Berkeley and Los Angeles: University of California Press, 2000)

For early medical thinking, see:

Luis Garcia Ballester, Roger French, Jon Arrizabalaga and Andrew Cunningham, *Practical Medicine from Salerno to the Black Death* (New York: Cambridge University Press, 1994)

Faye Marie Getz, *Medicine in the English Middle Ages* (Princeton, NJ: Princeton University Press, *c.* 1998)

V. Nutton, 'Humoralism', in W. F. Bynum and Roy Porter (eds.), *Companion Encyclopedia of the History of Medicine* (London: Routledge, 1993), 281–91

Nancy G. Siraisi, *Medieval and Early Renaissance Medicine: An Introduction to Knowledge and Practice* (Chicago and London: Chicago University Press, 1990)

For anatomy and physiology, see:

Andrea Carlino, *Books of the Body: Anatomical Ritual and Renaissance Learning*, trans. by John Tedeschi and Anne C. Tedeschi (Chicago: University of Chicago Press, 2000)

Allen G. Debus, *The Chemical Philosophy: Paracelsian Science and Medicine in the Sixteenth and Seventeenth Centuries* (New York: Science History Publications, 1977)

Robert G. Frank, *Harvey and the Oxford Physiologists: Scientific Ideas and Social Interaction* (Berkeley: University of California Press, 1980)

Roger French, 'The Anatomical Tradition', in W. F. Bynum and Roy Porter (eds.), *Companion Encyclopedia of the History of Medicine* (London: Routledge, 1993), 81–101

—, *William Harvey's Natural Philosophy* (Cambridge: Cambridge University Press, 1994)

Lester S. King, *The Medical World of the Eighteenth Century* (Chicago: University of Chicago Press, 1958)

—, *The Philosophy of Medicine: The Early Eighteenth Century* (Cambridge, MA: Harvard University Press, 1978)

C. D. O'Malley, *Andreas Vesalius of Brussels 1514–1564* (California: University of California Press, 1964)

Ruth Richardson, *Death, Dissection and the Destitute* (London: Routledge & Kegan Paul, 1987)

K. B. Roberts and J. D. W. Tomlinson, *The Fabric of the Body: European Traditions of Anatomical Illustration* (Oxford and New York: Oxford University Press, 1992)

B. Schultz, *Art and Anatomy in Renaissance Italy* (Ann Arbor: UMI Research Press, 1985)

Pathology is handled in:

Saul Jarcho (trans. and ed.), *The Clinical Consultations of Giambattista Morgagni* (Boston: Countway Library of Medicine, 1984)

Russell C. Maulitz, *Morbid Appearances: The Anatomy of Pathology in the Early Nineteenth Century* (Cambridge and New York: Cambridge University Press, 1987)

CHAPTER FOUR: The Laboratory

For hospital medicine, see:

Erwin H. Ackerknecht, *Medicine at the Paris Hospital, 1794–1848* (Baltimore: Johns Hopkins University Press, 1967)

M. Foucault, *The Birth of the Clinic*, tr. by A. M. Sheridan Smith (London: Tavistock, 1973)

Nineteenth-century experimental medicine is covered in:

Thomas D. Brock, *Robert Koch: A Life in Medicine and Bacteriology* (Madison, WI: Science Tech Publishers, 1988)

—, *Justus von Liebig: The Chemical Gatekeeper* (Cambridge: Cambridge University Press, 1997)

W. F. Bynum, *Science and the Practice of Medicine in the Nineteenth Century* (New York: Cambridge University Press, 1994)

Gerald L. Geison, *The Private Science of Louis Pasteur* (Princeton: Princeton University Press, 1995)

Frederic L. Holmes, *Claude Bernard and Animal Chemistry: The Emergence of a Scientist* (Cambridge, MA: Harvard University Press, 1974)

Wesley W. Spink, *Infectious Diseases: Prevention and Treatment in the Nineteenth and Twentieth Centuries* (Folkestone: Dawson, 1978)

For the twentieth century, consult:

Michael Bliss, *The Discovery of Insulin* (Edinburgh: Paul Harris, 1983)

K. J. Carpenter, 'Nutritional Diseases', in W. F. Bynum and Roy Porter (eds.), *Companion Encyclopedia of the History of Medicine* (London: Routledge, 1993), 463–82

Roger Cooter and John Pickstone (eds.), *Medicine in the Twentieth Century* (Amsterdam: Harwood, 2000)

Horace Judson, *The Eighth Day of Creation: Makers of the Revolution in Biology* (New York: Simon and Schuster, 1979)

Daniel J. Kevles and Leroy Hood (eds.), *The Code of Codes: Scientific and Social Issues in the Human Genome Project* (Cambridge, MA and London: Harvard University Press, 1992)

Gina Kolata, *Clone: The Road to Dolly and the Path Ahead* (London: Allen Lane, 1997)

Pauline M. H. Mazumdar, *Species and Specificity: An Interpretation of the History of Immunology* (Cambridge: Cambridge University Press, 1995)

J. C. Medvei, *A History of Clinical Endocrinology* (Lancaster: MTP Press, 1982)

Tom Wilkie, *Perilous Knowledge: The Human Genome Project and its Implications* (Berkeley: University of California Press, 1993)

Good on tropical diseases are:

W. D. Foster, *A History of Parasitology* (Edinburgh: E. & S. Livingstone, 1965)

Gordon A. Harrison, *Mosquitoes, Malaria, and Man: A History of the Hostilities since 1880* (New York: Dutton, 1978)

Sheldon Watts, *Epidemics and History: Disease, Power and Imperialism* (New Haven and London: Yale University Press, 1997)

Christopher Wills, *Yellow Fever, Black Goddess: The Coevolution of People and Plagues* (Reading, MA: Addison-Wesley Pub., 1996)

M. Worboys, 'Tropical Diseases', in W. F. Bynum and Roy Porter (eds.), *Companion Encyclopedia of the History of Medicine* (London: Routledge, 1993), 511–60

CHAPTER FIVE: Therapies

General studies:

E. H. Ackerknecht, *Therapeutics from the Primitives to the 20th Century* (New York: Hafner, 1973)

C. D. Leake, *An Historical Account of Pharmacology to the 20th Century* (Springfield, IL: C. C. Thomas, 1975)

Miles Weatherall, 'Drug Therapies', in W. F. Bynum and Roy Porter (eds.), *Companion Encyclopedia of the History of Medicine* (London: Routledge, 1993), 911–34

Specific topics in chronological order of subject:

John M. Riddle, *Dioscorides on Pharmacy and Medicine* (Austin, TX: University of Texas Press, 1985)

J. Worth Estes, *Dictionary of Protopharmacology: Therapeutic Practices, 1700–1850* (Canton, MA: Science History Publications/ Watson Publishing International, 1990)

Leslie G. Matthews, *History of Pharmacy in Britain* (Edinburgh and London: E. & S. Livingstone, 1962)

M. Weatherall, *In Search of a Cure: A History of Pharmaceutical Discovery* (Oxford; New York: Oxford University Press, 1990)

John Harley Warner, *The Therapeutic Perspective. Medical Practice, Knowledge and Identity in America, 1820–1885* (Cambridge, MA: Harvard University Press, 1986)

Nancy Tomes, *The Gospel of Germs: Men, Women and the Microbe in American Life* (Cambridge, MA: Harvard University Press, 1998)

J. Liebenau, *Medical Science and Medical Industry: The Formation of the American Pharmaceutical Industry* (London: Macmillan, 1987)

Ronald Hare, *The Birth of Penicillin and the Disarming of the Microbe* (London: George Allen and Unwin, 1970)

Arabella Melville and Colin Johnson, *Cured to Death: The Effects of Prescription Drugs* (London: Secker and Warburg, 1982)

Lara Marks, *Sexual Chemistry: A History of the Contraceptive Pill* (New Haven and London: Yale University Press, 2001)

Edward Shorter, *A History of Psychiatry. From the Era of the Asylum to the Age of Prozac* (New York: Wiley, 1997) [strong on psychopharmacology]

Useful on drug innovations is:

James Lefanu, *The Rise and Fall of Modern Medicine* (London: Little, Brown, 1999)

CHAPTER SIX: Surgery

Robert Bud, *The Uses of Life: A History of Biotechnology* (Cambridge/New York: Cambridge University Press, 1994)

Renée C. Fox and Judith P. Swazey, *The Courage to Fail: A Social View of Organ Transplants and Dialysis*, 2nd ed. (Chicago: University of Chicago Press, 1978)

Knut Haeger, *The Illustrated History of Surgery* (New York: Bell, 1988)

Ghislaine Lawrence, 'Surgery (Traditional)', in W. F. Bynum and Roy Porter (eds.), *Companion Encyclopedia of the History of Medicine* (London: Routledge, 1993), 957–79

Mark M. Ravitch, *A Century of Surgery: 1880–1980*, 2 vols (Philadelphia: J. B. Lippincott Co., 1982)

Ira M. Rutkow, *Surgery: An Illustrated History* (St Louis: Mosby-Year Book Inc., in collaboration with Norman Pub., 1993)

Ulrich Tröhler, 'Surgery (Modern)', in W. F. Bynum and Roy Porter (eds.), *Companion Encyclopedia of the History of Medicine* (London: Routledge, 1993), 980–1023.

Anthony F. Wallace, *The Progress of Plastic Surgery: An Introductory History* (Oxford: William A. Meeuws, 1982)

Owen H. and Sarah D. Wangensteen, *The Rise of Surgery: From Empiric Craft to Scientific Discipline* (Minneapolis: University of Minnesota Press, 1978; Folkestone, Kent: Dawson, 1978)

The problem of sepsis is the subject of:

Irvine Loudon, *Death in Childbirth: An International Study of Maternal Care and Maternal Mortality 1800–1950* (Oxford: Clarendon Press, 1992)

For childbirth and obstetrics, see:

Jacques Gélis, *History of Childbirth: Fertility, Pregnancy and Birth in Early Modern Europe* (Oxford: Polity Press, 1991)

Michael J. O'Dowd and Elliot E. Philipp, *The History of Obstetrics and Gynaecology* (New York and London: The Parthenon Publishing Group, 1994)

Adrian Wilson, *The Making of Man-Midwifery: Childbirth in England 1660–1770* (London: University College Press, 1995)

CHAPTER SEVEN: The Hospital

Broad surveys:

Lindsay Granshaw and Roy Porter (eds.), *The Hospital in History* (London and New York: Routledge, 1989).

Guenter Risse, *Mending Bodies, Saving Souls: A History of Hospitals* (New York: Oxford University Press, 2000)

J. D. Thompson and G. Goldin, *The Hospital: A Social and Architectural History* (New Haven and London: Yale University Press, 1975)

Specific studies in chronological order of subject:

T. S. Miller, *The Birth of the Hospital in the Byzantine Empire* (Baltimore: The Johns Hopkins University Press, 1985)

Nicholas Orme and Margaret Webster, *The English Hospital, 1070–1570* (New Haven and London: Yale University Press, 1995)

Colin Jones, *The Charitable Imperative: Hospitals and Nursing in Ancien Régime and Revolutionary France* (London and New York: Routledge, 1990)

Christine Stevenson, *Medicine and Magnificence: British Hospital and Asylum Architecture 1660–1815* (New Haven and London: Yale University Press, 2000)

J. Woodward, *To Do the Sick No Harm. A Study of the British Voluntary Hospital System to 1875* (London and Boston: Routledge & Kegan Paul, 1974)

Charles E. Rosenberg, *The Care of Strangers: The Rise of America's Hospital System* (New York: Basic Books, 1988)

Rosemary Stevens, *In Sickness and in Wealth: American Hospitals in the Twentieth Century* (New York: Basic Books, 1989)

Joel D. Howell, *Technology in the Hospital: Transforming Patient Care in the Early Twentieth Century* (Baltimore: Johns Hopkins University Press, 1995)

Stanley Joel Reiser, *Medicine and the Reign of Technology* (Cambridge: Cambridge University Press, 1981)

For clinical science, see:

Christopher Booth, 'Clinical Research', in W. F. Bynum and Roy Porter (eds.), *Companion Encyclopedia of the History of Medicine* (London: Routledge, 1993), 205–29

A. McGehee Harvey, *Science at the Bedside: Clinical Research in American Medicine 1905–1945* (Baltimore: Johns Hopkins University Press, 1981)

David Weatherall, *Science and the Quiet Art: Medical Research and Patient Care* (Oxford: Oxford University Press, 1995)

On nursing, see:

Monica E. Baly, *Florence Nightingale and the Nursing Legacy* (London: Routledge, 1988)

Susan M. Reverby, *Ordered to Care: The Dilemma of American Nursing 1850–1945* (Cambridge: Cambridge University Press, 1987)

The mental hospital is briefly covered, with reading suggestions, in Roy Porter, *Madness: A Brief History* (Oxford: Oxford University Press, 2002)

CHAPTER EIGHT: Medicine in Modern Society

For the critique of modern medicine, see:

Ivan Illich, *Limits to Medicine: Medical Nemesis: The Expropriation of Health* (Harmondsworth: Penguin, 1977)

Lynn Payer, *Disease-Mongers: How Doctors, Drug Companies, and Insurers are Making You Feel Sick* (New York, et al.: John Wiley & Sons, 1992)

For public health and state medicine, see:

Peter Baldwin, *Contagion and the State in Europe 1830–1930* (Cambridge: Cambridge University Press, 1999)

Carlo Cipolla, *Public Health and the Medical Profession in the Renaissance* (Cambridge: Cambridge University Press, 1976)

John Duffy, *The Sanitarians: A History of American Public Health* (Urbana and Chicago: University of Illinois Press, 1990)

Anne Hardy, *The Epidemic Streets: Infectious Disease and the Rise of Preventive Medicine, 1856–1900* (Oxford and New York: Oxford University Press, 1993)

Dorothy Porter, *Health, Civilization and the State* (London: Routledge, 1999)

George Rosen, *A History of Public Health* (New York: M.D. Publications, 1958; new edition, ed. by Elizabeth Fee, with updated bibliography by Edward T. Morman, Baltimore: Johns Hopkins University Press, 1993)

The medical profession is analysed in:

Jeffrey L. Berlant, *Profession and Monopoly: A Study of Medicine in the United States and Great Britain* (Berkeley: University of California Press, 1975)

P. Starr, *The Social Transformation of American Medicine* (New York: Basic Books, 1982)

For modern medicine and medical policy, see:

David Armstrong, *Political Anatomy of the Body: Medical Knowledge in Britain in the Twentieth Century* (Cambridge: Cambridge University Press, 1983)

Roger Cooter and John Pickstone (eds.), *Medicine in the Twentieth Century* (Amsterdam: Harwood, 2000)

Daniel M. Fox, *Health Policies, Health Politics: British and American Experience 1911–1965* (Princeton: Princeton University Press, 1986)

Daniel M. Fox, 'The Medical Institutions and the State', in W. F. Bynum and Roy Porter (eds.), *Companion Encyclopedia of the History of Medicine* (London: Routledge, 1993), 1196–222

Derek Fraser, *The Evolution of the British Welfare State: The History of Social Policy Since the Industrial Revolution* (London: Macmillan, 1973)

Anne Hardy, *Health and Medicine in Britain since 1860* (Basingstoke: Palgrave, 2001)

Helen Jones, *Health and Society in Twentieth-Century Britain* (London and New York: Longman, 1994)
Charles Webster, *The National Health Service: A Political History* (Oxford: Oxford University Press, 1988)

Totalitarian abuse of medicine forms the subject of:

Benno Müller-Hill, *Murderous Science: Elimination by Scientific Selection of Jews, Gypsies and Others in Germany 1933–1945* (Oxford: Oxford University Press, 1988)
Robert N. Proctor, *Racial Hygiene: Medicine Under the Nazis* (Cambridge, MA and London: Harvard University Press, 1988)

For medicine's present and future problems, see:

Laurie Garrett, *Betrayal of Trust: The Collapse of Global Public Health* (New York: Hyperion, 2000)
William L. Kissick, *Medicine's Dilemmas: Infinite Needs versus Finite Resources* (New Haven: Yale University Press, 1994)
James Lefanu, *The Rise and Fall of Modern Medicine* (London: Little, Brown, 1999)
T. McKeown, *The Role of Medicine: Dream, Mirage or Nemesis?* (Princeton: Princeton University Press, 1979)

Index

Page references in italic indicate illustrations; those given in bold indicate a definition or an explanation.